Systems Analysis and Management:

Structure, Strategy and Design

Systems Analysis and Management:

STRUCTURE, STRATEGY AND DESIGN

Donald V. Steward
California State University, Sacramento

PBI

a petrocelli
book

new york / princeton

Printed in the United States

1 2 3 4 5 6 7 8 9 10

Book Design: RFS GRAPHIC DESIGN, INC.

Library of Congress Cataloging in Publication Data

Steward, Donald V
    System analysis and management.

    Bibliography: p.
    Includes index.
    1. System analysis.   2.   Systems engineering.
I. Title.
T57.6.S72   658.4′032   80-15395
ISBN 0-89433-106-X

SYSTEMS ANALYST'S PRAYER

May I be granted:

1. The ease that comes from solving sequentially that which can be,

2. The strength to solve simultaneously that which must be,

3. And the wisdom to know the difference.

Contents

Preface

Preface

Some years ago, while working on the design of a nuclear power plant, we[1] realized that by rearranging an iterative computing scheme we could obtain a better design in a few computer minutes than we had previously been able to obtain after many computer hours. This was a significant insight at the time. It occurred to me that we might study the structure of systems and the flow of information during their design so that we could arrive at such insights in a more systematic way.

My early paper on this subject [DVS:62] came to the attention of an econometrician[2] who recognized the application of the idea to the simulation and analysis of models of the economy. Then a chemical engineer[3] recognized that these techniques could be applied to chemical engineering process design. These methods have also been applied to the system of differential equations describing the control of a rocket.[4] Where we had started by looking at just one facet of a design, these methods are now being used to look at the management of the whole design process (see chapter 7). Now similar techniques have been applied to societal problems.[5]

[1]Wesley Harker and the author, at General Electric.
[2]Professor Charles Holt, then at the University of Wisconsin.
[3]Professor Dale Rudd, Chemical Engineering Department, University of Wisconsin.
[4]Peter Benyon, Commonwealth Scientific and Industrial Research Organization, Canberra, Australia.
[5]John Warfield, *Societal Systems* (New York: John Wiley, 1976).

Systems can be described by their structure and semantics. The structure is represented by a graph or matrix showing which parts affect what other parts. The semantics concern how the effects occur. We will show how to represent and analyze the structure of a system to plan a strategy for its study or design.

The planning and scheduling of the fabrication of a system is fundamentally different than the planning and scheduling of the design of that same system. The fabrication of a system does not involve circuits and thus can be planned and scheduled by the now familiar critical path methods. But the design of a system does involve circuits. These circuits represent simultaneous relationships that must be resolved before the fabrication can begin. Circuits in a design process are typically handled by starting with a guess to make a preliminary design, then iterating to obtain improved designs. Since circuits are not tolerated by the standard critical path techniques, it has become obvious that a new method for planning and scheduling the design of systems is needed. We present such a method here.

The tools we present help one plan where assumptions are to be used, how design iterations occur, where design reviews are required, and what are the durations of the tasks for each iteration. Once these decisions are made, we can lay out the design process as a finite sequence of activities which can be scheduled and controlled by the usual critical path techniques.

Algebra allowed us to move beyond the limits of simple word problems solvable in our heads to more complex problems. This we did by representing the structure of these problems with a formalism which led us along a series of easily comprehended steps to the final solution. Here we present a system algebra which allows us to analyze or design large systems. Again we describe the structure of the system by a formalism, and use this description to develop a strategy for the solution or design of the system in simple, comprehendible steps.

Many of society's problems have become too complex to handle without making severe simplifications. Experts tend to make different simplifications, which can lead them into conflict about what is the real problem and its solution. We will discuss ways in which we can help the experts to put their individual pieces of the system puzzle together so that instead of each expert feeling just a trunk, a leg, or a tail, they can all see and agree that they are dealing with an elephant (see chapter 9).

Gabriel Kron developed a method for solving large systems of linear equations by "tearing." He broke large systems into smaller subsystems, which he then solved subject to constraints representing the interfaces with the other subsystems. From these solutions he constructed the solution of the whole system. Only a few people who had Kron's ability to see where and how to tear have had any real success in applying his methods. We have developed a different, more general concept of tearing and have provided a tool that helps the user see where to tear systems.

In chapters 1 through 5 we develop methods for representing and studying the structure of a system in order to develop strategies for studying, designing, or managing that system.

Chapters 6 through 9 discuss several applications of these techniques. Chapter 6 sets the necessary background in precedence methods for critical path planning, scheduling, and resource allocation; several new methods not previously published are introduced. Chapter 7 applies the techniques to planning, analysis, and scheduling of the engineering design of complex systems. Here the tasks are the determination of the design variables, the execution of complex computer codes or design procedures, and the preparation of documents. Chapter 8 considers the application of these techniques to the analysis and diagnosis of econometric models represented by systems of equations and to econometric input-output models. Chapter 9 discusses the elephant problem—how to put together the various facets of a system as seen by individual experts, to see and manage the whole system.

The structure of a system is represented by a graph showing which parts are affected by which other parts. The vertices of the graph represent the parts of the system. There is a path from vertex x_i to x_j if and only if the behavior of x_i affects the behavior of x_j. The semantics of the system concerns the rules for the behavior of the parts and their effects on each other. The operation of determining or designing the behavior of any one part, given the assumed or determined behavior of the parts which affect it, is treated as a task.

If we number the parts in the order in which we propose to solve or design them, a high-numbered part affecting a low-numbered part is a feedback. If there are circuits in the graph, then there must be some such feedback. By whatever means we choose to solve or design a system, feedbacks will play a special role, e.g., where guesses are made to begin iteration, or where variables are carried symbolically during elimination, etc. This book considers the structure of the system and how it affects the relation between this ordering and the resulting feedbacks. This leads us to a strategy for the study, design, or management of the system.

Computer techniques are developed which (1) "partition" the tasks into blocks such that circuits exist only between tasks in the same block, and (2) analyze these blocks to generate what we will call shunt diagrams. Users then employ these shunt diagrams plus their knowledge of the semantics of the system to select "tear" arcs whose removal would leave the graph with no circuits. The tasks are renumbered so that these tear arcs or a subset become the feedbacks. This numbering and the resulting feedbacks then define the strategy for determining or analyzing the behavior of the system.

Chapter 2 introduces the concepts of partitioning and tearing. Several applications derived from distinctly different technologies are presented to motivate these concepts and demonstrate the generality of the techniques. Methods for generating and using shunt diagrams for selecting tears are presented in chapter 4.

Partitioning and tearing are also applicable to systems of n equations in n unknowns, provided an assignment is made of a dependent variable in each equation. These techniques are considered in chapter 5 and are used in chapter 8.

Readers will want to read this book for different purposes. Some will want to know only the fundamental ideas—how to formulate the problems to be solved by these methods and how to interpret the results. Others will want to understand the methods in such detail that they are able to carry out the methods and even extend them. Some won't be interested in the application of these methods to systems of equations, while others will. The table of contents and Figure A guide the readers in establishing their own tour through the book to suit their needs.

Figure A shows which chapters should be read in order to understand each chapter. For example, to understand the application to managing the engineering design of systems (chapter 7), one should first read chapter 2, and if one were not already familiar with precedence methods for critical path scheduling, one should also read chapter 6. Those who are reading chapter 2 only as a prerequisite to the applications in chapters 7, 8, and 9 need not dwell on some of the more difficult concepts, but should merely scan the chapter. Chapters 5, 7, and 8 contain sections at the end which use tearing. Anyone interested in reading the sections of these chapters on tearing should first read chapter 4 and look at the Appendix which discusses the output of the TERABL program used in the illustrations.

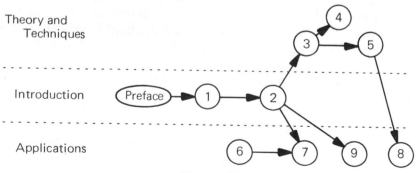

Figure A
A Guide to Reading This Book

I Introduction to Systems

1.1 What Is a System?

The term *system*, or phrase *complex system*, appears more and more frequently in the technical literature. We read about management systems, computer systems, ecological systems, endocrine systems, hospital systems, information systems, space flight systems, etc. What is a system and why is it such an important and fundamental concept that it is used extensively in so many different fields?

We like to define systems by the following characteristic: "A system is a thing which if you make any change to it there are likely to be many subtle consequences." Although many might contest the precision of this definition, it does capture the essence of our concern about systems, namely, it is important to understand the consequences of changes, which may not be immediately obvious, before we dare make the changes.

The more classic definition of a system, based on a model of how we perceive the system might be constructed, is as follows: "A system is a collection of parts and relations between the parts such that the behavior of the whole is a function not only of the behaviors of the parts, but also of the relations among them."

A part may itself be a system, in which case we call it a *subsystem*. By *behavior* we mean the rules that describe the relation between the input (what is done to it) and the output (how it responds). Note that by this definition, *behavior* concerns the rules which relate how the part responds to any input, not just what it is doing now.

1.2 Analysis and Complexity

To understand systems we often analyze them, which literally means breaking them into parts (more properly, this form of analysis is called reduction). The expectation is that (1) the individual parts by themselves will be easier to understand, and (2) by understanding the behavior of each of the parts and how they interact, we can then infer and thus understand the behavior of the whole. This leads to the following definition of complexity: "Given the parts and their behaviors, complexity is the difficulty involved in using the relations among the parts to infer the behavior of the whole." Or phrased another way, complexity is how much more the whole is than just the sum of its parts.

1.3 Complexity and Growth of Technology

The recent growth in technology has been primarily in the dimension of complexity. We are not now advancing our technology by the application of new, previously unknown laws so much as by the greater complexity with which we assemble parts, each of which behaves according to well-established laws. Even as new laws become known, they are discovered by the increased ability to analyze more complex systems. Most of the simple systems have been analyzed or designed long ago.

Thus, it is reasonable to conclude that a significant contribution to the advancement of science and technology would come from an increased ability to analyze more complex systems.

Computers have made a vital contribution to our capabilities for handling complex systems, as witness the accomplishments of the space program, the development of nuclear reactors, etc., which would not have been possible without computers. But we do not always understand the results of computer calculations well enough to generalize and expand on them, nor do we always know how to ask the computer to process complex data so that we can understand the results. Even though we have computers, we still need the help of additional methods to extend our capabilities to understand complex systems.

Not only are technological systems becoming more complex, but so are social and political systems. The greater reach of interactions between persons and between societies stemming from more complex transportation and communication systems, and, unfortunately, from more complex weapons, has made people and societies more interdependent and their systems more complex. As we begin to see that we are reaching the final limits of our resources, again an effect accelerated by the growth of complex technology, the allocation of these remaining resources places further interactive pressures on society. Complexity breeds complexity. The problems of solving and/or resolving complex systems problems is a key not only to our technological growth, but to our very survival.

1.4 Systems Problems

More and more of the problems we face are systems problems. Systems problems, by definition, cannot be solved by considering one part or aspect of the problem at a time without also considering how this part affects or is affected by other parts. All too often, rather than face this reality, we approach systems like the proverbial blind men approached the elephant. Each expert sees the problem from his own perspective, concentrating on one facet as though it alone represented the whole system. Arguments occur as to which part of the elephant is the true elephant. Is the leg, which appears to be a tree, the true elephant? Or the tail, which appears to be a rope, or the trunk, which appears to be a hose, the really true elephant? Is the urban transit problem one of providing faster transportation or one of more convenient access, or of more reliable or comfortable transportation? Arguments can easily ensue unless a model is developed which shows the structure of how all these aspects interrelate and contribute to some specified goal.

The increase in complexity of society and technology means that we are faced with making decisions about more complex systems. Decisions are often required quickly to keep things moving whether or not there is time to make the analysis needed to properly understand all the possible consequences. When a decision is made, to be comfortable, the decision maker must feel that there is a rationalization for the decisions, i.e., that he had a basis from which his decision followed. Thus there is a tendency to grab at any available simplistic approach to rationalize and defend decisions. Even our electoral process is based upon politicians presenting to the voters overly simplistic solutions and rationalizations for what are really very complex systems problems.

If management can afford the time and thinking space to consider the problem in the required depth, the systems analyst would be wise to involve management in deriving the solutions. But this happy circumstance is the exception rather than the rule.

The systems analyst must usually present his proposed solutions to management through heavily taxed communications channels. He must make his points quickly and precisely and bring management to a position where they can see a comfortable rationalization to support the proposed solution. The systems analyst is also responsible to feel comfortable in his belief that he has presented the best solution considering management's needs and his own, more comprehensive understanding of the system.

1.5 Systems Integration

It often appears that the easiest way to integrate the understanding of and responsibility for complex systems decisions is to have all the decisions centralized in one person's head. Some individuals have an unbelievable capacity to integrate large, complex systems. However, as the systems get larger, more complex and more extensive, an organization which has functioned well under this scheme may find there comes a point when this one key person becomes overburdened. At this stage the system has not been planned for decentralization of system responsibility and authority. The organization now has all its facilities taxed to their limits contending with the day-to-day problems. Under these circumstances, the organization may sink into chaos before a careful management plan for decentralized decision making can be accomplished.

A means is required to analyze the information flow inherent in the system which the organization deals with in order to develop an organization with delegated, decentralized responsibility and authority responsive to problem solving in that system environment. The organization chart should represent a strategy for solving a systems problem by a decomposition of the problem into components involving separated responsibility and authority, as well as communications between components. This organization chart should reflect the information flow revealed by an analysis of the structure of the system.

Information flow analysis is also vital to the design of distributed systems, or what might better be called distributed problem-solving systems. In such systems a problem or need is resolved by a network of interconnected computers, data bases and people. The information flow analysis developed here may be used to allocate responsibilities for these parts of the system.

1.6 Decomposition

How do we go about breaking a system into parts so that we might understand it? Clearly, one way to break up a system is with an explosive. However, we would have great problems trying to understand the original system by studying the parts we could collect after the blast. First, these parts would not represent the most meaningful units to study. Second, the relations among the parts, which are by definition important to understanding the behavior of a system, would have been destroyed.

To understand a system we must take it apart carefully. To take it apart, we must sever the relations between the parts. But as we do so, we must study those relations if we are to understand how the system as a whole works. We may later test the hypothesis that we understand these relations by putting the parts back together in their proper relationships so that the system as a whole behaves as it did originally.

As we break a system into smaller and smaller parts, the parts usually become easier to understand. But the number of relations among this greater number of parts increases, which makes more difficult the problem of assembling the understanding of the parts into an understanding of the whole. Just how difficult depends upon how many, which, and how the relations were severed in decomposing the system. Thus, we need methods to choose how to make these divisions.

Often we are not able, or not willing for ethical reasons, to physically take a system apart and put it back together. It is inappropriate to dismember certain societies or the human body to gain an understanding of how they behave. Furthermore, we do not have the confidence that, having dismembered them, we will be able to understand them well enough to put them back together in their original working form.

1.7 Models

When we cannot break the real system into parts, we make a model that we can break into parts. A "model" is a *system* and a *relation* between the model and the real system, so that understanding of the model can be translated, using these relations, into an understanding of the real system. The model may be mathematical, where the relation might be that a count of certain things in the model corresponds to a count of corresponding things in the real system.

The modeler's art requires great skill and judgment. The model should be simpler and easier to understand than the real system, otherwise, we might not be able to understand our model either. The model should be a good representation of the real system, for otherwise our understanding of the model would not contribute to an understanding of the real system. Furthermore, unless the behavior is sufficiently similar to the behavior of the real system, it might be very difficult to infer the relation between the behaviors of the model and the real system, and how the model may be improved to make a better model.

Thus, the art of modeling involves developing a model which is simple, yet sufficiently represents the important behavioral characteristics of the real system so that properties of its behavior can be recognized and used to make improvements in the model.

The model is a hypothesis of how we believe the system is built. The exercise of the model to obtain conditions and behavior which can be related to the real system is called *simulation*. By simulation we test the model and compare its behavior to its real world counterpart. In engineering we may be able to control the tests of the real system. In social systems we usually cannot make experiments, so we may have to use experiments which have already occurred naturally.

In the early formative stages, when the model is crude and incomplete, it may be extremely difficult to make a comparison between the behavior of the model and the real system. As the model is developed and becomes a more accurate representation, these comparisons become easier.

How to develop models, compare their behavior to that of the real system, and from that comparison diagnose and cure the weakness in the model is a very sophisticated art. It is the very art of science itself. Once we have a model which behaves sufficiently like the real system, it is hoped we can understand the model and by this means understand the real system.

When we break a system into parts, then break those parts into parts, etc., we produce a hierarchial decomposition of the system; see Figure 1.1. This diagram is called a *tree*. The tree fans cut as each part is broken into smaller parts. Most systems can be broken into parts in any of a number of ways. Choices of how to break down the system can have different consequences on how difficult it is to study the parts and their interactions to obtain an understanding of the whole. Often we need to study the relations among the parts before we can determine how to make this decomposition.

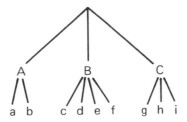

Figure 1.1
Hierarchical Decomposition

1.8 Structure and Graphs

The tree shows how the system is decomposed into parts, but it does not show the relations among those parts. Whether this particular hierarchial decomposition represents a good decomposition depends upon how the parts defined by the decomposition interact. We use another tool, called a *graph,* to represent how the parts are connected to form the system.

The term graph as used here is a generally accepted term for a set of points, called *vertices,* and directed lines between points called *arcs.* We do not care where the vertices are placed on the paper or how the arcs are drawn between them, only which vertices are connected to which other vertices by the arcs. This use of the term graph should not be confused with the more common meaning of the term to represent a plot of one variable as a function of another; Figure 1.2 shows this type of graph. Figure 1.3 shows the type of graph that we will be concerned with here.

We use the vertices of a graph to represent the parts of a system, and

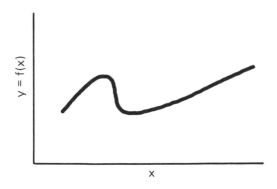

Figure 1.2
Graph—Meaning Plot

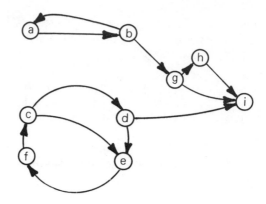

Figure 1.3
Graph—As It Will Be Considered in This Book

the arcs to show which parts affect which others. An arc is drawn from one vertex to another to show that the behavior of the one part affects the behavior of the other part.

Thus a graph shows which parts of a system affect which other parts. This we call the *structure* of the system. In other words, the structure shows where the parts are connected, or related. The graph does not show how or why the one part affects the other, only whether the one part affects the other. We call this "how and why the effects occur" the *semantics* of the system. Together, the structure and semantics completely describe the system.

The structure of a system may be that A and B each affect C; see Figure 1.4. The semantics for this structure could be drawn from econometrics and say that the Agricultural (A) and Manufacturing (B) sectors contribute to the GNP (C). Or, semantics drawn from engineering could mean that we can determine the size of the motor (C) given the weight (A) and the aerodynamic drag (B). Or the semantics could be drawn from an urban model, in which case it would mean that poor public

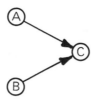

Figure 1.4
Structure, which Could Have Different Semantics

protection (A) and poor property maintenance (B) contribute to vandalism (C). In all these examples the structure is the same, but the semantics are different.

The distinction between structure and semantics used here is analogous to the distinction between syntax and semantics in the study of language. Syntax is the structure of a sentence described by such terms as "subject" and "predicate," while semantics concerns what the sentence means.

We make this distinction between structure and semantics in order to develop techniques whereby we can abstract and study the structure of a system, no matter what its semantics, to develop a strategy for the study or design of that system. This analysis of the structure does not depend upon whether the semantics of the system comes from econometrics, engineering, biology, management theory, etc. The same techniques can be used whether the semantics concerns the operations of a hospital, the design of a power plant, the study of the economy, or whatever.

1.9 Matrices

Another way of representing the structure of a system is with a matrix. For example, the graph in Figure 1.3 can be represented by the matrix in Figure 1.5. The rows and the columns correspond to the vertices in the graph.

	a	b	c	d	e	f	g	h	i
a	⊗	X							
b	X	⊗							
c			⊗			X			
d			X	⊗					
e			X	X	⊗				
f					X	⊗			
g		X					⊗		
h							X	⊗	
i				X			X	X	⊗

Figure 1.5
Matrix of Graph in Fig. 1.3

Each non-blank mark in the matrix corresponds to an arc in the graph. A mark in column i row j represents an arc from vertex i to vertex j.

The essential aspect of a system, that aspect which makes it a system rather than just a collection of parts, is its structure. Thus, any study or design of a system, other than very trivial systems, must consider the structure of the system. As the systems we deal with become more complex, we need more formal tools to describe and analyze system structures. The most natural tool to describe this structure is a graph, or its equivalent, the matrix.

We must first analyze the structure of a complex system to give us a road map that will then guide us as we analyze the semantics of the system.

Now, with these descriptive tools, the graph and its corresponding matrix, we can begin to reconsider in more precise terms some of the concepts discussed above. The next chapter does this.

2 Systems and Their Structure

2.1 Introduction

This book will consider methods for analyzing how the parts of a system interact and how information flows in the system. From this analysis we shall obtain systematic plans for analyzing the behavior of the system, or for managing the design of the system. These concepts can be applied to any type of system. To illustrate the breadth of applications, we shall consider as examples the process of designing an electric car, and the analysis of a model of the United States economy represented by a system of simultaneous equations. Other systems to which these techniques have been applied include chemical process design, analysis of the control system for a missile, and data flow in a data processing system.

Remember that a system is a set of parts such that the behavior or design of a part may depend upon the behavior or design of other parts. We shall describe a system by examining its structure and its semantics. The structure of the system is a description of what parts affect what other parts. The semantics of the system is how the behavior or design of a part is determined once the behavior or design of the parts which affect it are known. This is sometimes called the technology of the system.

This book is concerned primarily with the analysis of the structure of the system to plan its design or study its behavior. Some examples are presented which imply specific types of semantics, e.g., the sequencing of design tasks, the solution of equations, the flow of money, etc. However, it would be impossible to consider here all the types of semantics to which these techniques could be applied. It is assumed that the reader is already familiar with the semantics of the specific systems he wishes to study. We shall show how to analyze the structure of the system so as to plan and manage the strategy for its design or analysis.

2.2 Representing the Structure of Systems

The first technique for representing the structure of a system is a *graph*. Graph theory is a well-established discipline dating back at least to Euler (circa 1736). We shall not assume that the reader has any previous familiarity with graph theory; everything will be developed from first principles. An established graph theory terminology will be used so that those familiar with graph theory can relate to what we do here.

A graph is a set of vertices (points) and a set of arcs (directed lines) between certain pairs of vertices. The vertices represent the parts of the system. The arcs are drawn from one vertex to another vertex that it affects. In Figure 2.1a the arrow from vertex 1 to vertex 2 implies that part 1 affects part 2. A *path* is a sequence of one or more arcs from vertex to vertex in the direction of the arrows, such as the path (1,2,4) in the figure. An arc shows a *direct effect,* such as part 1 directly affects part 2. A path of

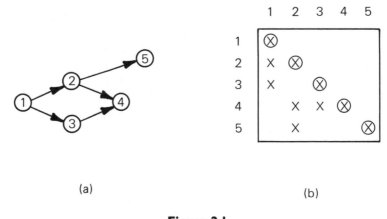

(a)   (b)

Figure 2.1
(a) Graph with no circuits. (b) Precedence matrix for graph in part (a).

two or more arcs, such as (1,2,5), shows *indirect effects*. The *length of the path* is the number of arcs in the path. The length of the path (1,2,5) is thus 2. If there is a path from *a* to *b*, *a* is said to be a *predecessor* of *b*, and *b* is a *successor* of *a*.

A central concern of our analysis of structure is the circuit. A *circuit* is a path of more than one arc that leads back to the starting vertex. Figure 2.1a has no circuit. Figure 2.2a has several circuits, such as (2,3,2), (6,9,6), and (6,2,3,9,6).

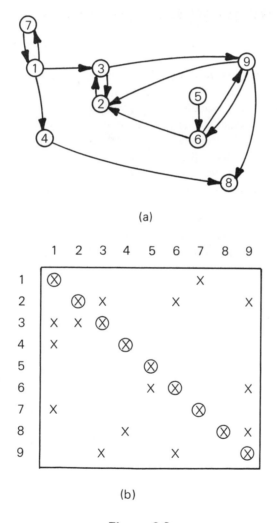

(a)

(b)

Figure 2.2
(a) Graph with circuits. (b) Precedence matrix for graph in part (a).

Another method of representing the structure of a system, which is equivalent to a graph, is the precedence matrix. A *precedence matrix* is a square matrix with as many rows and as many columns as there are vertrices in the graph. The element m_{ij} (row i, column j) contains a mark if and only if there is an arc to vertex i from vertex j. By convention, we also put circled x's on the diagonal. Figures 2.1b and 2.2b show the precedence matrices corresponding to the graphs in figures 2.1a and 2.2a, respectively.

If the number of vertices and arcs is small, it is easy to work with a graph. However, as the number of vertices and arcs increases, a precedence matrix becomes easier to work with than a graph. Furthermore, a precedence matrix can be input to a computer for help with the analysis.

The techniques discussed here apply interchangeably to either the study of the behavior of a system or the design of a system to have a specified behavior. Rather than mention both aspects with every reference, we may mention just one, leaving it to the reader to observe that the statement applies equally well to the other.

We consider in this book how to analyze the graph or precedence matrix in order to formulate plans for the analysis or design of the system. This plan will show the order in which we consider the parts, where estimates must be used for information we do not yet have, how to iterate the analysis or design of subsystems, where design reviews are required and what must be reviewed, and how the design process should be managed. It will also show how the resources are scheduled and controlled, what is affected by changes, and how changes are controlled, documented and verified.

2.3 Simultaneous Subsystems

In our analysis of structure, numbers are assigned to the vertices of the graph. These numbers represent the order in which we consider the parts in our analysis or design plan.

We assume that we can determine the behavior or design of any part, given just the behaviors or designs of the parts that directly affect it. We might like to find a numbering so there is no arc from a high-numbered vertex to a low-numbered vertex. Then, proceeding in the order of these numbers, we could analyze or design one part at a time. As each part is considered, the behavior or design of all the parts which affect it would have lower numbers and thus have already been determined.

Can we always obtain such an ordering? The next chapter shows that we can if and only if the graph has no circuits.

A circuit corresponds to a simultaneous subsystem. Simultaneous subsystems can be solved either by special methods such as solving simultaneous equations by elimination, or they can be solved by iteration. To solve by iteration, we begin the first iteration by making a guess for some variable in the circuit. We say that this breaks the circuit. Then we proceed, solving for each variable in turn around the circuit, until we get back to solving for the variable we originally guessed. Now this new solution gives a new guess. We can decide whether our original guess was adequate, or use the new guess to make another iteration.

Given a graph and a numbering of the vertices, an arc from a high-numbered vertex to a low-number vertex is called a *feedback*. For any numbering, if there is a circuit, at least one arc in each circuit must be a feedback.

If we break one arc in each circuit, then the graph that is left has no circuits. Then it can be numbered so that there is no feedback. When we restore the torn arcs, they occur as feedbacks and appear above the diagonal in the matrix.

If our plan calls for consideration of the parts of a system in the order of the numbers assigned to the vertices, then a feedback represents where we must assume the behavior of a higher numbered part we have not gotten to yet; that is, if we are able to handle the parts in a circuit by iteration, the feedbacks show where we need to make guesses to begin the iteration. A feedback arc from a to b implies we guess at a while we determine b.

In the solution of a system of simultaneous equations, feedback arcs represent variables carried algebraically during the elimination until they can be solved for later. (We discuss systems of simultaneous equations in chapter 5.)

2.4 Partitioning and Tearing

We break the process of ordering into two phases. Phase 1 is called partitioning. Phase 2 is tearing.

Partitioning divides the vertices into blocks. These blocks are defined such that (1) within each block there are paths in both directions from every vertex in the block to every other vertex in the same block, and (2) between a vertex in one block and a vertex in another block, there can be at most a path in one direction. This defines the blocks uniquely. The blocks represent simultaneous subsystems, i.e., each part is simultaneously dependent upon all the other parts of the same block.

Numbers are assigned to the vertices such that (1) the vertices within a block are assigned contiguous numbers, and (2) no arc goes from a high-numbered vertex in one block to a low-numbered vertex in another block.

Partitioning breaks the system into blocks representing subsystems. Within each block there is a path from every vertex in the block to every other vertex in the same block. These blocks are uniquely defined by the structure, shown by the graph or precedence matrix.

Although the partitioning assigns a series of numbers to order the vertices of each block relative to the vertices of other blocks, it does not say anything about the relative ordering of the vertices within the block. Initially, the vertices within the blocks are arbitrarily left in the same order as they occurred before the partitioning.

Parts (a) and (b) of Figure 2.3 show the partition of the system in parts (a) and (b) of Figure 2.2, respectively.

In the second phase, *tearing,* we consider one block at a time to determine how the vertices should be relatively ordered within the block. Note that a reordering of the vertices within a block will affect which marks occur above the diagonal. Marks above the diagonal correspond to feedbacks. Parts (c), (d), (e), and (f) of Figure 2.3 show other possible orderings of the vertices within the blocks.

The choice of ordering the vertices within each block is based upon evaluation of the semantics of the feedbacks which occur with that

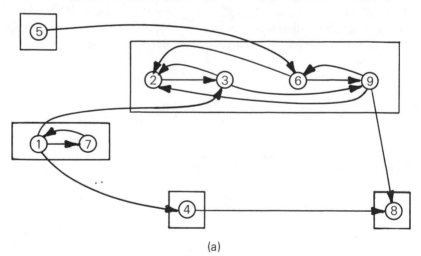

(a)

Figure 2.3

(a) Partition of the graph in Fig. 2.2a. (b) Partitioned matrix for Fig. 2.2b; matrix corresponding to Fig. 2.3a. (c) Partition of the graph in Fig. 2.2a with another ordering of the vertices within the blocks (obtained from the tears shown) ⟵+⟞ shows tears. (d) Partitioned matrix for Fig. 2.2b, corresponding to Fig. 2.3c. (e) Partition of the graph of Fig. 2.2a with yet another tearing and its ordering. (f) Matrix corresponding to Fig. 2.3e.

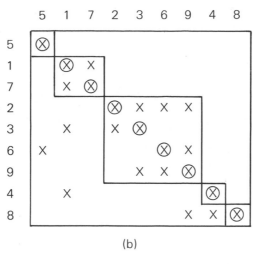

(b)

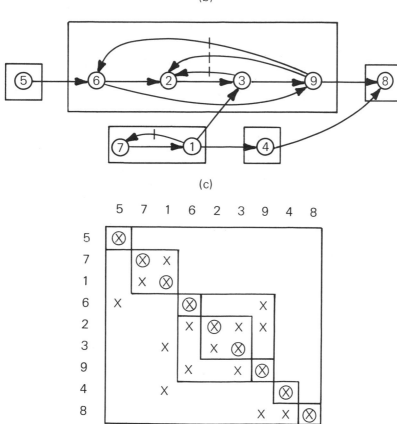

(c)

(d)

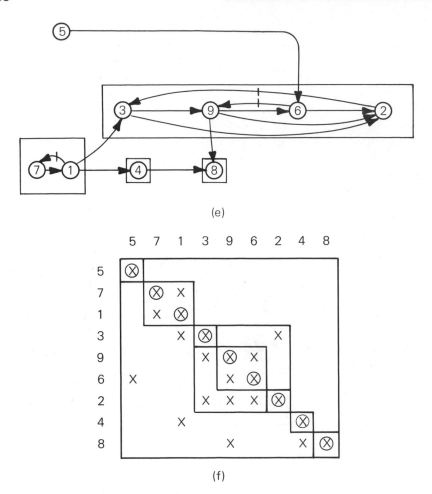

(e)

(f)

ordering. No matter what semantics we are dealing with in the system, whether it is economic input-output relations, differential equations, mass balance, or whatever, a feedback requires some form of special consideration. If the system is to be solved iteratively, the feedbacks represent where guesses are used to initiate the iteration. If the system is a set of simultaneous equations to be solved by elimination, feedbacks represent variables carried as unknowns during the elimination process.

By whichever means we solve a simultaneous system or subsystem, there is a special role played by the ordering and resulting feedback arcs. They represent the carrying of an unknown or the making of a guess. Feedback arcs increase the cost of the solution. But, depending upon the semantics and the method used to solve or design the system, there may be significant differences in the costs of the various orderings and their feedbacks. It is not necessarily true that the ordering with the least number of feedback arcs is the least costly. We must consider the semantics of the feedbacks before making a final judgment of which of several possible orderings gives the best strategy for the analysis or design of the system.

There are two approaches that can be taken to finding an ordering of the vertices within a block to obtain the most desirable (or, more properly, the least undesirable) set of feedbacks.

First, we could try an ordering. Then the feedbacks are the arcs (or marks) from high-numbered vertices to low-numbered vertices in this ordering. We consider the semantics to evaluate the costs of making guesses for these feedbacks.

Second, we could begin by considering the semantics and choose arcs where we can easily afford to make the required guesses. We choose these arcs so there is one such arc in each circuit. Then we tear these arcs. If we ignore these torn arcs, then no circuits remain. Now we can reorder the vertices within the block by our partitioning procedure.

We usually prefer the second approach because it lets us work from our evaluation of the arcs as possible feedbacks (more likely to be a known) to the ordering (which is as yet unknown).

Assume we use approach two. If after removing a set of arcs and partitioning, any blocks having more that one vertex remain, we can then remove one or more arcs from each of these remaining blocks, reorder the vertices within these smaller blocks, etc., until no nontrivial blocks (blocks with more than one vertex) remain. When no nontrivial blocks remain, we know we have removed sufficient marks that all the circuits have been broken by some tear.

If after removing a set of marks and partitioning we find a nontrivial block, we could also begin again with another set of marks to be torn and repeat the whole process.

This process is called tearing because removing a mark from the matrix corresponds to tearing an arc in the graph. The idea is to tear a set of arcs so that all the circuits are broken. Then partitioning the vertices within the block produces an ordering of the vertices within the block.

A later chapter on tearing explains how the use of shunt diagrams shows those sets of tears which meet certain necessary conditions for them to break all the circuits. Then only the arcs in one of these sets need be evaluated to determine whether they represent acceptable feedbacks.

2.5 Applications and Semantics

In the design of a system, arcs may represent information flow. The vertices may represent design tasks that can be done using the information from the predecessor tasks. A design task could be the determination of a design variable, the running of a computer program or design procedure, or the writing of a document. We speak interchangeably about the vertex in the graph or what the vertex represents in that application; a part of a system, a task, or a design variable. For example, a vertex may represent the area of a heat exchanger, or the task of determining that area.

In the fabrication of systems, arcs may represent enabling conditions, while the vertices represent tasks. Cement can be poured once forms are in place and the cement is mixed. In an input-output model of the economy, the vertices represent products or sectors of the economy and the arcs represent the flow of dollars. In a system of equations a vertex represents an equation and the dependent variable it is solved for, while the arcs show which variables appear as independent variables in the equation. (However, the application of these techniques to systems of equations introduces some new considerations, which are discussed in chapter 5.)

Critical path schedules (CPS) may be represented by graphs. The vertices represent activities and the arcs represent constraints on the order in which the activities can be done. (As we shall see in the chapter on critical path scheduling, other representations are also possible, as, for example, the conventional PERT chart with the activities on the arcs.) In the conventional use of critical path scheduling the graph represents a time sequence. A circuit in a critical path schedule would imply that an activity is constrained not to begin until it has already been finished. This is an absurdity. A restriction on CPS graphs is that they not contain a circuit. This is no problem when CPS is used to plan and control the fabrication of a system because we do not expect to have circuits during fabrication. But it is a problem in planning the design of a system because, as we shall see, if our graph were to show the relation between tasks or variables in the design of a system, it would contain circuits.

For example, in the design of a heat exchanger, the temperature depends upon the heat flux, the heat flux depends upon the heat transfer coefficient, and the heat transfer coefficient depends upon the temperature. This is a circuit. The engineer or system designer must resolve this circuit before the heat exchanger can be built. He may do this by iteration. He guesses the temperature to determine the heat transfer coefficient, which he then uses to determine the heat flux; the heat flux is used to reestimate the temperature. Or he may set up a system of simultaneous equations which he solves for the whole set of variables. Once a $100,000 heat exchanger is installed, it would be very expensive to find it was too small and, worse yet, to find there was no room to install a larger one.

An engineer's or systems designer's role is to resolve the circuits during the design phase so that no circuits occur in the fabrication. It is less expensive to resolve these circuits with pencil and paper, a computer calculation or simulation, or a test on a prototype or similar existing system than it is to use trial and error in building the full-scale system.

The remainder of this chapter considers the analysis of two distinctly different types of systems. The first is the design of an electric car, a project that involves many people, with many disciplines, working on different subsystems. The concern of management is that during the design the people work together successfully, and, when the system is built, that the subsystems work together successfully.

The second example is the analysis of a model of the economy. It illustrates the application of our techniques to systems of equations.

2.6 The Design of an Electric Car

Although we shall be a bit naive in the simplicity we assume about the semantics of electric cars, we shall, nevertheless, illustrate the basic techniques.

We begin by listing the pertinent tasks in the design; see Table 2.1. These tasks can be the determination of design variables such as the size of the electric motors, or the execution of computer codes or design procedures, or the preparation of documents.

Each task is assigned a number for identification. Next to each task in the table we list by number the predecessor tasks. The predecessor tasks are those that must be completed or their results estimated before this task can be done. We have listed the predecessors in two columns. In the first column are the sensitive predecessors. We say that a predecessor is *sensitive* if the likely error in the estimate of the predecessor could have a big effect on the task it precedes. A predecessor is *insensitive* if we can make a good estimate, or if a bad estimate would not have a big effect. Choosing whether an effect is sensitive or insensitive, or how insensitive, is a matter of technical judgment.

Figure 2.4 shows this same information in the form of a precedence matrix. The rows and columns are numbered by the tasks in our table. A mark in row i column j indicates task i is preceded by task j. Zeros indicate the sensitive predecessors, and nines indicate insensitive predecessors. We could distinguish more levels of sensitivity if we wish.

Imagine a triangular matrix where all the marks are either on or below the diagonal. Then we could perform each task one at a time in the sequence they are listed in the matrix. As we look at the row representing a task, all its predecessor marks appear to the left of the diagonal, which means they have already been done.

TABLE 2.1 Precedence Table

| | | Predecessor Tasks | |
| | | More Sensitive | Less Sensitive |
Variables	Task Description	(0)	(9)
1	Passenger capacity specifications	None	
2	Size–aerodynamics	1,7	3,11,12
3	Motor specifications and weight	2,4,11	6,7
4	Total weight	1,2,7,12	3,11
5	Stored energy requirement	8,9,13	3,6,10
6	Battery type–energy density	None	
7	Battery size and weight	5,6	
8	Cruising speed specifications	None	
9	Speed and acceleration performance vs. power	2,4	1,12
10	Acceleration specifications	None	
11	Speed and acceleration conformance	8,9,10	
12	Structural and suspension design	4	1,2,3,7,11
13	Range specification	None	
14	Cost	2,3,4,6,7,12	
15	Consumer demand vs. cost	1,8,10,13	2,12
16	Profit	14,15	

Can we always reorder the tasks, making the same reordering of both rows and columns, so that the matrix comes out triangular? If there is a circuit, no. Given an ordering of the rows and columns in the matrix, the marks above the diagonal are arcs from high-numbered vertices to low-numbered vertices. Thus they represent feedbacks. If there are circuits, then for any ordering at least one arc in each circuit must be a feedback and occur as a mark above the diagonal.

If there are circuits, although we cannot obtain an ordering that will make it triangular, we can obtain an ordering that makes the matrix block triangular with square blocks on the diagonal. These blocks on the diagonal are the blocks of our partition. They are the subsystems in which all of the parts are interrelated.

In Figure 2.5 we have partitioned and reordered the tasks in Figure 2.4. Note that this is a block triangular matrix. The matrix is triangular except for square blocks on the diagonal. Most of these blocks contain only one task. There is also a block of eight tasks. We cannot find any ordering for this block where all the marks are on or below the diagonal. Notice that we can trace the following sequence of predecessors that gives a circuit: 2 is preceded by 7 (the zero in row 2, column 7), 7 is preceded by 5, 5 is preceded by 9, 9 is preceded 4, and 4 is preceded by 2. Now we are back to the beginning of the circuit.

Precedence matrix (rows and columns numbered 1–16):

	1	2	3	4	5	6	7	8	9	10	11	12	13	14	15	16
1	⊗															
2	0	⊗									9	9				
3		9	⊗			9	9				9					
4	0	0	9	⊗			0				9	0				
5			9		⊗	9		0	0	9			0			
6					9	⊗	0	0								
7					0		⊗	0								
8								⊗	0	9						
9	9		0						⊗			9				
10			0						0	⊗	9					
11										0	⊗	9				
12	9	9	0					0	9			⊗	9			
13	0	0											⊗			
14	0	0			0	0		0		0				⊗	0	
15	0	9									9		0		⊗	
16													0	0		⊗

Row labels:
1. Passenger capacity specifications
2. Size–aerodynamics
3. Motor specifications and weight
4. Total weight
5. Stored energy requirement
6. Battery type–energy density
7. Battery size and weight
8. Cruising speed specifications
9. Speed and acceleration performance vs. power
10. Acceleration specifications
11. Speed and acceleration conformance
12. Structural and suspension design
13. Range specifications
14. Cost
15. Consumer demand vs. cost
16. Profit

Figure 2.4

Precedence Matrix with Level Numbers for the Design of an Electric Car. 9's Mark A Priori Tears.

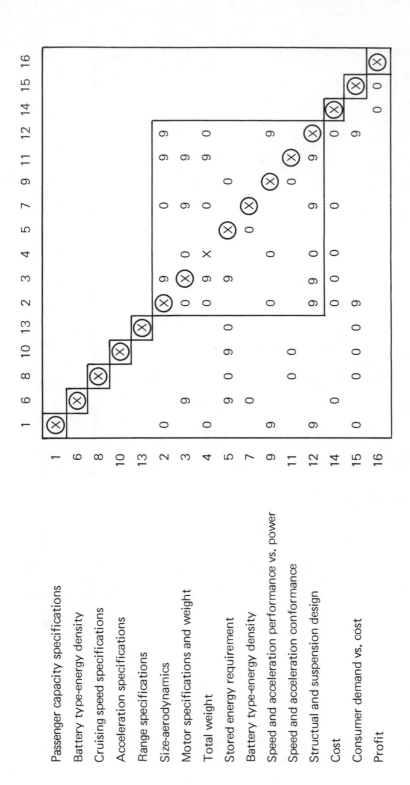

Figure 2.5
First Partition—Using All Levels

The ordering of the tasks within the block will establish which marks occur above the diagonal. These marks are the feedbacks that represent where we must use estimates for the information we will not have available until later. A mark in row i column j implies we must guess at variable j to determine variable i. The ordering, where we use estimates and when we have the data to review these estimates, is the basis for our design strategy.

We might do the tasks in the order they occur in the block in Figure 2.5. We would start by using guesses for the results of tasks every place a mark occurs above the diagonal. The column shows which variable was estimated, while the row shows where the estimate is used. When we get to the end of the block, we will have made a determination of the results of each task, based upon these first guesses. We could then have a design review to compare these new values with our guesses. If they agree well enough to satisfy us, we would be finished with the block. If not, we would use our new values to make better guesses and repeat the process. This could be continued until we are confident we have a consistent set of values that satisfy the whole system.

If we order the tasks in the block differently, we get different marks above the diagonal. These new marks represent different sets of estimates. We may have a better basis for making these new estimates and, as a result, the process may converge more rapidly. So we must consider how the tasks within the blocks should be ordered, the effects this ordering has on the estimates to be made, and how the process converges. This depends upon both the structure and the semantics. So, to partition we need consider only the precedence matrix or graph. But to order the tasks within the blocks we must use our understanding of the semantics as well as the structure.

We would like to have an ordering within each block such that we have good estimates, or can afford to use poor estimates, for the marks above the diagonal. In this example we used the number 9 for those marks where we can see right off that we can afford to use estimates. Note that these marks initially may be either above or below the diagonal. We can test to see whether it is possible to get an order within the block with only the 9's above the diagonal by the process of tearing. We remove the 9's from the block, then partition the block. Having obtained a new order in the block by this partition, we now restore the torn marks to their proper rows and columns. This produces the reordered matrix in Figure 2.6. We have used the term *tearing* because we removed marks from the matrix, which corresponds to tearing arcs in the graph. (We shall explain shortly why those two 0 marks were changed to 5's.)

We note that we were not successful in getting only 9's above the diagonal. So we must make some additional tears, or find another set of tears.

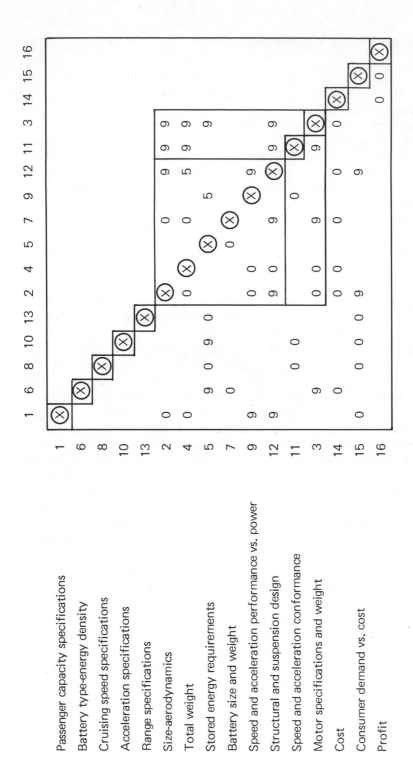

Figure 2.6

Partition of Remaining Block—Ignoring 9's, 5's Represent A Posteriori Tears

26

Let us draw a box around the block containing the zeros remaining after our last partition. Now we look for choices of zero marks to tear that would break all the circuits. If we tear all the circuits, then when we partition, all the remaining marks would be on or below the diagonal. Only the marks we tore, or a subset, would remain above the diagonal.

One such set of marks to tear in this example would be the marks in row 5 column 9, and row 4 column 12. These are the marks we changed to 5's in Figure 2.6. Other choices might be to tear (7,9) (row 7 column 9) and (12,4), or tear (2,7), (4,7) and (12,4). There are a number of possibilities; we would choose among them by bringing to bear our knowledge of the semantics. We must consider what it would mean to the design process to make the estimates these tears represent.

When working with large systems we can use a computer to perform the partitioning process. The next chapter describes a procedure that can be done either by hand or on the computer. The computer would use numbers we assign to the marks for the order in which it will tear them. The highest numbered marks are removed first, and the block is partitioned. Then the next highest numbers are removed, etc., until the only remaining marks are all zeros. We have used 9's to indicate the first set of marks we want torn, and 5's to indicate the next set to tear. We must use our knowledge of the semantics and our judgment to assign the numbers. If after removing all the non-zero marks and repartitioning there still remain blocks with more than one vertex per block, the computer program will print a shunt diagram. This diagram will help us see what combinations of tears of the remaining marks will break all the circuits. (Shunt diagrams are discussed in the chapter on tearing. The TERABL computer program is discussed in the appendix.)

The process of tearing is an interaction between the computer and the user. The computer will analyze the structure, remove the marks and repartition, then show by means of shunt diagrams what combinations of tears will break all the remaining circuits. With our knowledge of the semantics and our judgment, we can choose from these possibilities a set of tears that represents a reasonable design strategy. We tell the computer where we have chosen to tear by assigning numbers to the marks. By this interactive process with the computer, or by hand if the problem is a small one, we can work out an ordering and strategy for the design of a system.

Anytime we develop a plan for the design of a system, whether we use these techniques or just stumble through a design, we have implicitly made choices of how the system is torn and ordered. The methods of this book will not guarantee optimum choices, but will lead to ways to make better choices that will result in better designs with lower design costs.

No matter how we choose the ordering and design plan, there is a great advantage to making up the matrix and revealing in a clear way to everyone involved in the decision process what the design procedure is. Then each person can make his decisions consistent with the same plan. This plan will show what tasks will be affected by changes, where design reviews are required, and what should be reviewed. It will also show how the design work can be scheduled and controlled. There are many benefits just from using the matrix to communicate the plan clearly, no matter how the plan was derived.

By partitioning and tearing our simple electric car example, we have obtained the matrix in Figure 2.7. This represents a statement of the design process and is called the *design structure matrix*. In a common, mutually understandable language, it shows everyone involved in the design precisely where the guesses are made to begin the design process and where there should be design reviews. It indicates by the marks above the diagonal which variables or tasks should be compared to the estimates in that review.

It often occurs that a change is made in some variable during the design process due to finding an error or oversight, a new idea, or a change in the requirements. By following down columns one can trace the successors of the changed tasks, and in turn the successors of these tasks, etc., until all the consequences are traced. Then a plan for implementing the change can be made. Similarly, the matrix can be used to trace the tasks and develop a plan to modify a standard design to satisfy specific customer requirements.

The matrix becomes a key tool in change control and design verification. It shows the information flow involved in the design process and who must communicate with whom. This suggests how the organization should be set up and how the documents should be designed. (These matters are developed in more detail in chapter 7.)

Once we estimate how many times we expect to iterate each block and the duration of each task in each iteration, we have the input to a critical path schedule. Chapters 6 and 7 discuss how iterations can be unwrapped to obtain critical path schedules.

If we were to look at the fabrication of the system rather than the design, we would not expect to find circuits. The designer should resolve the circuits at design time so that there are no circuits at fabrication time.

The reader who is primarily interested in applications to designing systems such as we have discussed in this first example, and who wishes to know how to collect the data, formulate the problems, and use the results, can proceed from here directly to chapters 6 and 7, on critical path scheduling and the design structure system.

In the next chapter we consider the theory and the techniques used in partitioning systems whose structures are described by a graph or precedence matrix. Readers interested in the actual procedures, or who otherwise wish to understand the theory, should read that chapter.

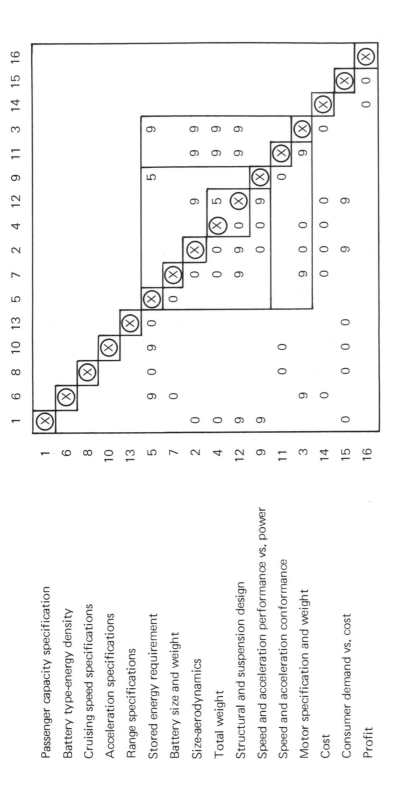

Figure 2.7
**Partition of Remaining Block—Ignoring 5's and 9's
Design Structure Matrix**

The theory of tearing and the use of shunt diagrams are more complex than most of the other material treated in this book. Therefore, to make it easier for the reader who does not wish to get involved in the details, we have deferred that theory to chapter 4; applications of tearing techniques to problems considered in the body of a chapter are presented at the end of that chapter.

2.7 Systems of Equations Representing an Econometric Model

We now apply partitioning and tearing to a system of n simultaneous equations to be solved for the values of n unknown variables.

Figure 2.8 depicts a matrix where the rows represent equations and the columns represent variables. The marks show where the variables appear in the equations. This example is the revised version of the Klein-Goldberger annual model of the United States economy [Klein and Goldberger:55].

Note that unlike figures 2.1 and 2.2 the rows and columns in Figure 2.8 represent different sets. This matrix maps the set of variables onto the set of equations. We circle one mark in each row so that one and only one circle occurs in each column. This assigns a unique dependent variable to each equation. We permute the rows so that the circles are brought to the diagonal, then label each row by the variable assigned to that equation. This gives Figure 2.9, which is a precedence matrix for a directed graph. It can be partitioned like the design problem we just considered. In Figure 2.10 the variables have been reordered by partitioning, revealing a square block of size 14 on the diagonal. This is a submodel.

An off-diagonal mark represents an independent variable in an equation. For a given ordering, a mark above the diagonal represents a feedback from an equation which has not yet been solved. It may be desirable while studying and improving the model to make a simplification by replacing these feedbacks with estimates for the variable obtained from outside the model. One might use as estimates actual historical values, a constant, a time series, or other simple model. This makes it possible to analyze the model one equation at a time to see whether each equation behaves properly. If the model designer had to consider all the equations simultaneously, he might not be able to pin the cause of a misbehavior of the model on the specific equation or equations that caused the error.

We seek tears such that (1) the behavior of the model is not sensitive to replacing variables by estimates every place a mark is torn, and (2) the

Variables

C	I	S_p	P_c	D	W_1	N_w	w	F_I	A_1	p_A	L_1	L_2	i_L	i_S	Y	P	p	K	B

- C — Consumer expenditures in 1939 dollars
- I — Gross private domestic capital formation in 1939 dollars
- S_p — Deflated corporate savings
- P_c — Deflated corporate profits
- D — Capital consumption in 1939 dollars
- W_1 — Deflated private employee compensation
- N_w — Number of wage and salary earners
- w — Index of hourly wages (1939 base: 122.1)
- F_I — Imports of goods and services in 1939 dollars
- A_1 — Deflated farm income excluding government payments for farmers
- p_A — Index of agricultural prices (1939 base: 100)
- L_1 — Deflated end-of-year liquid assets held by persons
- L_2 — Deflated end-of-year liquid assets held by enterprises
- i_L — Average yield on corporate bonds
- i_S — Average yield on short-term commercial paper
- Y — Deflated national income
- P — Deflated nonwage nonfarm income
- p — Price index of gross national product
- K — End-of-year stock of private capital in 1939 dollars
- B — Deflated end-of-year corporate surplus

Equations (⊗ = circled X, output assignment)

Eq.	C	I	S_p	P_c	D	W_1	N_w	w	F_I	A_1	p_A	L_1	L_2	i_L	i_S	Y	P	p	K	B
1			X			X				⊗	X					X	X			
2				X	X	X												⊗		
3														⊗						
4										⊗								X		
5			X			X			X			⊗		X		X				
6		X			X														⊗	
7	X	X			X			X							⊗					
8	⊗		X			X			X								X			
9		⊗																		
10					X								⊗	X				X		
11															⊗					
12					X				⊗	X								X	X	
13					X					X						X		⊗		
14				⊗													X			
15			⊗	X																
16					⊗											X			X	
17				X	⊗											X				
18							X	⊗												
19		X																		⊗
20			X				⊗									X		X		

Figure 2.8
Structural Matrix for Klein-Goldberger Model with Output Assignment

31

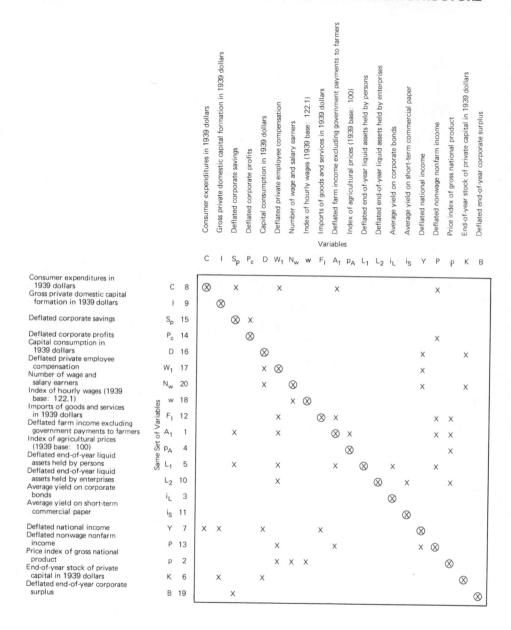

Figure 2.9
Klein-Goldberger Model with Rows Permuted to Bring Outputs to Diagonal
(Variables Are Now Assigned One to One to Equations)

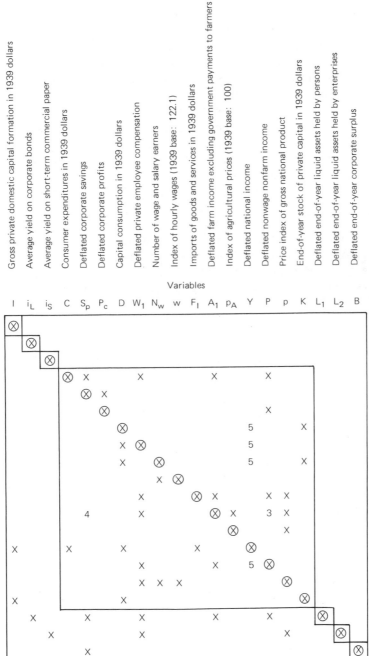

Figure 2.10
Klein-Goldberger Model—Partitioned and Level Numbers Assigned

remaining matrix when reordered has blocks representing meaningful submodels. Then, given the solutions of earlier submodels and estimates for the values of variables where the tears occur, each submodel can be studied independently. By this means submodels can be isolated to simplify finding the sources of errors in the behavior of the model so the model can be improved.

In Figure 2.11 several marks have been torn and the variables reordered to reveal a reasonable structuring of the model into submodels. If Y is assumed to be a known function of time, then the equations that use Y no longer depend upon equation 7. The model partitions into submodels as follows: a submodel involving four variables (A_1, p, P_c, S_p), a submodel involving two variables (D,K), and fourteen submodels of one variable each.

The same techniques used in these two examples may also be applied to systems involving flows and balances as in the design of chemical processing systems, or dollar flows and balances in econometric input-output models as illustrated in chapter 8. Although we discuss specific examples for concreteness, the reader will soon see that these principles may be applied to any type of system he may wish to consider.

2.8 Distributed Problem Solving

Computing has evolved through several stages in the way it has affected users. The last stage leads us into a revolution.

In the regime before the computer, people worked with their own files and processed transactions when and where they occurred. In the regime of batch computing, people gave over their files to other people to process. Then they waited for their results while transactions queued up to obtain batches large enough for economic processing on a large, centralized computer. Convenience, control, and timeliness were sacrificed to the economy of scale required to make computing economic.

As the costs and sizes of computers and terminals have come tumbling down, it is now cost effective to process transactions when and where they occur, directly under the control of the user of the information. Thus, the new regime looks to the user more similar to the precomputer regime, but he has increased his processing power enormously.

As computer storage costs have also come down and terminals are becoming available on everyone's desk, it becomes practical to store and distribute information through the computer without the use of paper. This is called *electronic paper*. Rather than each person having his own copy, each person can look at the same copy in computer storage.

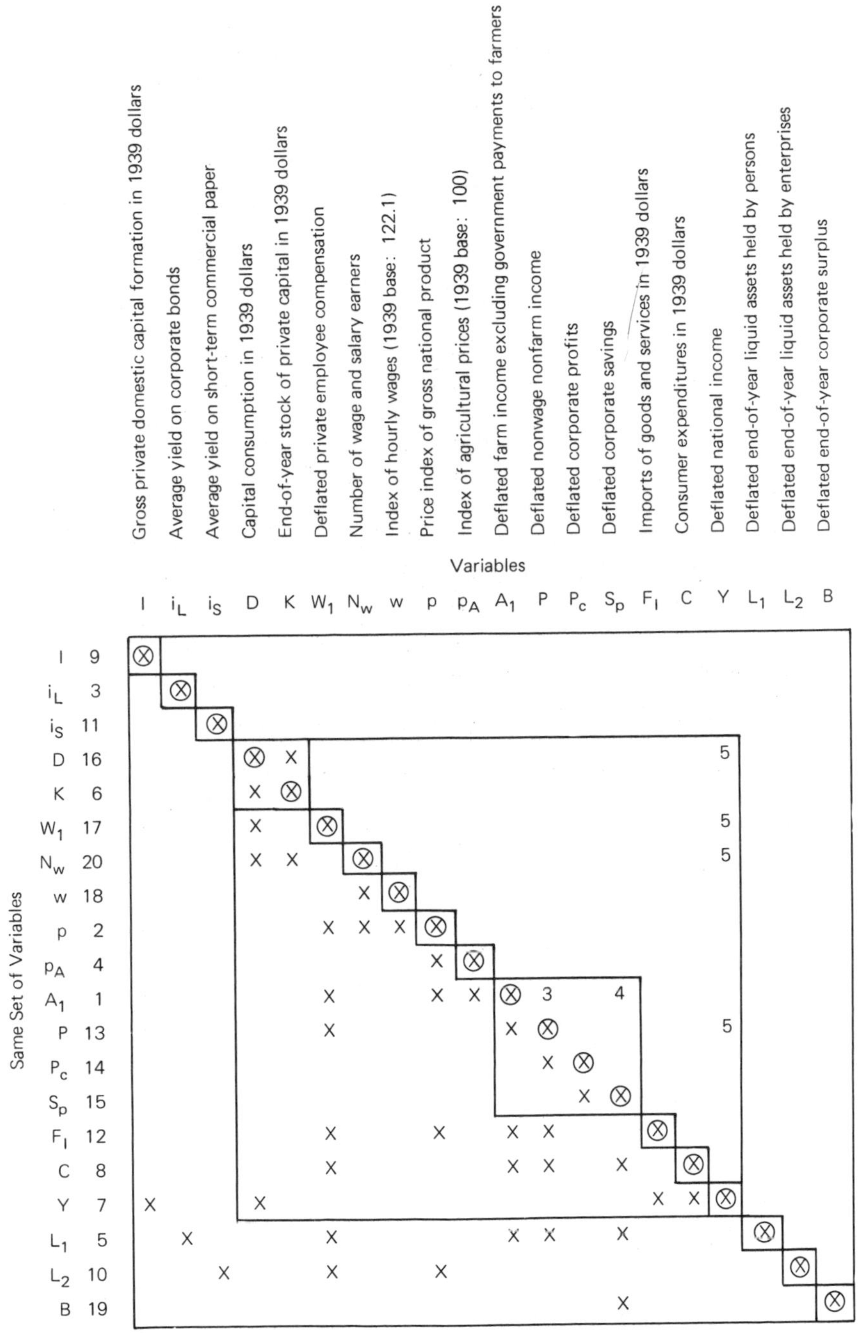

**Figure 2.11
Klein-Goldberger Model—Partitioned and Torn**

Now we have a system in which the terminal work station is the common interface by which to prepare input for the computer and look at the results, to send and receive information from fellow workers, to schedule and coordinate work, maintain records, etc. With his terminal he no longer needs many other office facilities such as a typewriter, filing cabinet, keypunches, copy machines, etc.

Now we have a problem-solving network of people and computers, all of whom you can talk to through the same terminal. Each person has a notification file which he checks regularly to see what information he needs has become available, what messages there are for him, what requests have been made by others for his services, what responses he has from requests he has made for the services of others, what tasks of his are about to become due, what meetings are scheduled, etc. These notifications refer him to files he can retrieve for more information. He can build a file of his tasks and priorities, and maintain his schedule. And he can monitor the state of the activities in the system.

This approach to distributed problem solving will require rethinking how we perform various old functions in the new scheme. We will see that while the design structure matrix can play a valuable role in making a paper-oriented system behave more efficiently, it plays an even more vital role in the problem-solving network using electronic paper.

The design structure matrix shows how the responsibilities for the tasks should be distributed among the partners, where information should reside, and who should be notified when it is available or changed. It is used in developing schedules and evaluating the effects of changes.

In a paper system, data is batched into documents for economic dissemination. This batching causes a time delay in making the information available. Electronic paper can accelerate the whole problem-solving process greatly by making information available as soon as it exists. But this raises problems and possibilities regarding how the validity of this information is controlled. When data is batched in documents, the document can be reviewed for validity before being released. But if a change occurs after the document is released, it is hard to catch and notify all those who have used the old data.

In an electronic paper system, notification of new information and access to the files containing that information can be controlled by access privileges. People verifying the data may be given access before it is released to others. Using electronic paper, a record can be made of everyone who has used the data so they can be informed of only those changes which occur to data they have used.

It will be during the lifetime of this book that the transitions to these methods will occur. Thus, chapter 7 has been written so that the techniques discussed can be implemented on a batch, timeshare, or distributed system. We expect the ideas developed in the book will be an invaluable aid in making the transition.

2.9 The Respective Roles of Graphs and Trees

Recent developments in the representation of the logic of computer programs raise some interesting questions about the respective domains of graphs and trees.

For many years flowcharts have been used to work out the logic of computer programs. Flowcharts are a form of graph. Iterations appear as circuits. Now it has been shown that computer program logic can be represented by trees.

Bohm and Jacopini (1966) have shown that the logic of any program can be represented using a hierarchical arrangement of three basic constructs: sequence, conditional (IF ___ THEN ___ ELSE ___) and iteration (DO WHILE). This hierarchy can be represented as a tree.

Sequence is represented by the order in which one traverses a tree. (Start at the left [or top] and go as far as you can. As you finish each branch, retreat until you can take the next branch to the right [or below].) The conditional is handled by entering a branch only if a condition on that branch is true. Iteration is handled by repeating the subtree entered from a branch as long as a condition on that branch is true. Figure 2.12 shows a flowchart and a tree for a program to compute standard deviation.

Trees have proved to be superior to flowcharts as a way of representing the logic of computer programs. Trees are easier to understand. They represent a map that allows one to work through the whole program in such a way that you only have to comprehend a small part at a time. This is in conformance with Miller's principle that says the human mind is quite limited in the number of items of information it can deal with at one time [Miller:56].

These concepts of writing computer programs are fundamental to the idea of structured programming. Warnier-Orr diagrams, which represent programs as trees, are replacing flowcharts about as rapidly as the word spreads.

All this raises a question: What other constructs now represented by graphs might be represented better by trees? The answer, we believe, lies in the distinction between description and prescription.

Description concerns how a system exists. The behavior of parts may be simultaneously interdependent. These simultaneous interdepen-

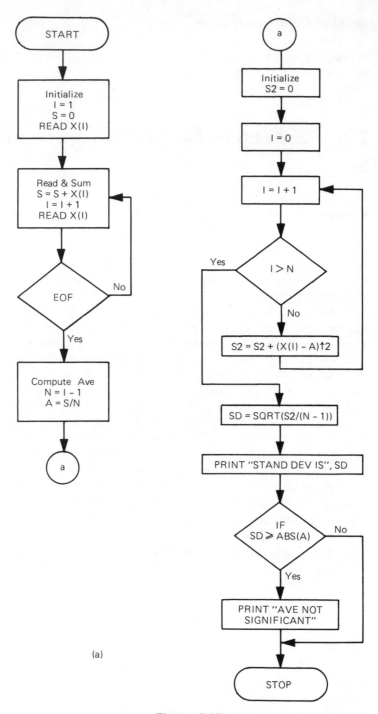

(a)

Figure 2.12
Flowchart and Tree for a Program to Compute Standard Deviation

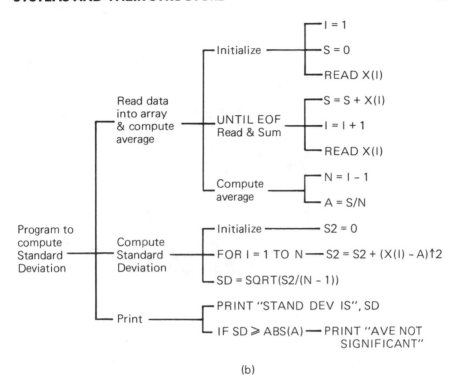

(b)

dencies appear as circuits in the cause and effect graph. *Description requires circuits.*

Prescription concerns how something is done, a procedure. A computer program is an example of a prescription. *Prescription can be represented by a tree.*

In this book wc discuss descriptions of systems we wish to exist or which already exist and we wish to change. These descriptions we represent by graphs or matrices which generally contain circuits. We then order the vertices to develop a prescription of how we proceed to build or change the system. The feed forward provides the procedure, which contains no circuits, while the feed back provides the control in the procedure. The ordering techniques described in this book transform the description of the system to be built or changed into the prescription, i.e., procedure, for doing the building or changing.

3 Theory and Techniques for Analyzing the Structure of Systems

3.1 Partitioning

In this chapter the concepts developed in chapter 2 are expressed in more formal mathematical language, and algorithms are shown for partitioning systems.

We consider the problem of ordering a set of vertices $x_1, \ldots, x_n \in X$ subject to a set of constraints. The vertices may represent tasks, and we thus use the terms *vertex* and *task* interchangeably. Tasks are activities which one intends to perform in some order, such as the solution or design of the behavior of one part of a system at a time. The constraints, i.e., one task should be done after, or depends upon, another, are represented by the arcs of a directed graph G.

DEFINITION   A *directed graph* $G = (X, U)$ is a finite set of elements $x_i \in X$ called *vertices* and a set of ordered pairs of vertices $(x_i, x_j) \in U$ called *arcs*. A *path* from x_i to x_j is a sequence of arcs such that the second vertex of each arc is identical to the first vertex of the next arc, except that the first vertex of the first arc is x_i and the second vertex of the last arc is x_j. A path is written as $(x_i, \ldots, x_j)$ where each adjacent pair of vertices corresponds to an arc in the path. The *length* of the path is the number of arcs in the sequence. A *circuit* is a path of length greater than one whose first and last vertices are the same and no proper subset is a circuit. A *subgraph* is a graph consisting of a subset of the vertices and those arcs between the vertices in this subset.

The vertices of the graph G are the tasks. A path from x_i to x_j implies that task x_i should precede (or affects) task x_j. Figure 3.1 shows a graph where vertices are represented by points and arcs are represented by directed lines between points. For example, (1,2,4) would represent the path from vertex 1 to vertex 2 to vertex 4. We seek to order these tasks by assigning a unique integer to each task such that there is no path from a high-numbered vertex to a low-numbered vertex. This can always be done by the following procedure [Marimont:59] provided the graph has no circuits.

PROCEDURE 3.1.   *Ordering of tasks in a graph without circuits*

Step 1: Set $I = 0$.

Step 2: Let $I = I + 1$. Since the graph (or any subgraph) contains no circuits, there exists a vertex with no predecessor, i.e., a vertex with no arc directed toward it. Remove such a vertex and all the arcs which exit from it. Number this vertex I. If any vertex remains, repeat Step 2. Otherwise, go to Step 3.

Step 3: This completes the procedure. Since every vertex was numbered after all its predecessors had been assigned lower numbers, there is no path from a high-numbered vertex to a low-numbered vertex.

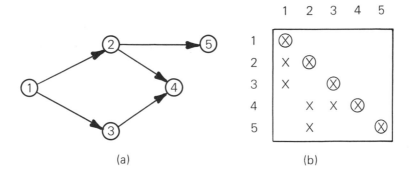

(a)   (b)

Figure 3.1
(a) Graph with no circuits. (b) Precedence matrix for part (a).

We are also concerned, however, with ordering tasks where the graph does contain circuits; see Figure 3.2. Clearly, it is not possible to number the vertices in a circuit such that there is no path from a high-numbered task to a low-numbered task. We are thus interested in orderings which must violate some of the constraints implied by the graph, i.e., some of the arcs must go from high-numbered vertices to low-numbered vertices.

DEFINITION   Given a graph and an assignment of distinct numbers to the vertices, an arc from a high-numbered vertex to a low-numbered vertex is called a *feedback arc*.

DEFINITION   *Partitioning,* as used here, means assigning the tasks to subsets called *blocks* and assigning distinct integers to the tasks such that the following conditions hold:

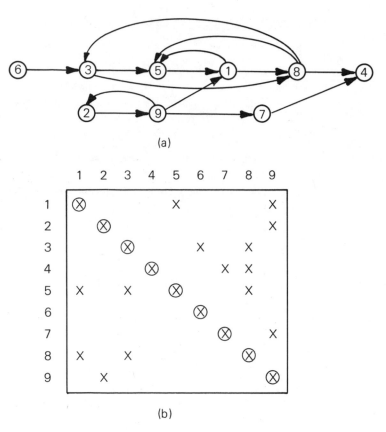

(a)

(b)

Figure 3.2
(a) Graph with circuits, e.g., (2,9,2) and (3,5,1,8,3). (b) Precedence matrix for part (a)

1. Tasks within a block are assigned contiguous integers.
2. No arc goes from a high-numbered task in one block to a low-numbered task in another block.
3. Given any two tasks x_i and x_j in the same block, there is a path from x_i to x_j and a path from x_j to x_i.

Figure 3.3 shows the partitioning of the graph in Figure 3.2.

We can also describe a block from two other perspectives. First, let us indicate that either there is a path from x_i to x_j or that x_i and x_j are identical by writing $x_i \leq x_j$. We then can see that the operator $\leq$ has the following properties:

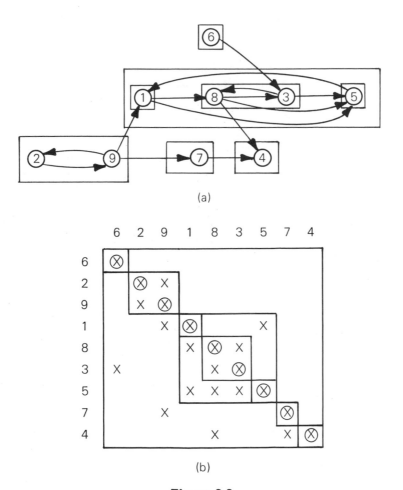

(a)

(b)

Figure 3.3
(a) Partition of the vertices in Fig. 3.2 into blocks. (b) Precedence matrix for part (a)

1) $x_i \leq x_i$ for all x_i (reflexivity)

2) $x_i \leq x_j$ and $x_j \leq x_k$ implies $x_i \leq x_k$ *(transitivity)*

If $x_i \leq x_j$ we say that x_i "precedes" x_j. Note that by definition each x_i precedes itself ($\leq$ produces a weak ordering). If we let $x_i \leq x_j$ and $x_j \leq x_i$ imply $x_i \equiv x_j$, then it follows that the operator $\equiv$ has the following properties:

1. $x_i \equiv x_i$ for all $x_i \in X$ *(reflexivity)*
2. $x_i \equiv x_j$ implies $x_j \equiv x_i$ *(symmetry)*
3. $x_i \equiv x_j$ and $x_j \equiv x_k$ implies $x_i \equiv x_k$ (transitivity)

and is called an *equivalence operator.* The equivalence operator partitions the tasks into equivalence classes, which are our blocks.

Second, since for any x_i and x_j in a block there is a path from x_i to x_j and a path from x_j to x_i, the graph theorist would say that each block is a strongly connected subgraph. (A graph is said to be *strongly connected* if for every pair of distinct vertices x_i and x_j there is a path from x_i to x_j and also a path from x_j to x_i.)

THEOREM 3.1.   *The blocks of the partition are disjoint and unique*. (This is a classical theorem for equivalence classes. See, for example, Berge:58.)

Proof: Assume B_k and B_l are two blocks of a partition with a task x in common. Then by transitivity and reflexivity every task in B_k and every task in B_l is equivalent to x and thus every task in B_k is equivalent to every task in B_l. Thus, if B_k and B_l are not disjoint they must be identical. If x_i and x_j are in the same block in any partition, then $x_i \equiv x_j$ and thus they must be in the same block in every partition. If x_i and x_j are in distinct blocks in any partition, then $x_i \not\equiv x_j$ and thus they must be in distinct blocks in every partition.

A finite directed graph can be represented by a precedence matrix as follows:

DEFINITION.   A *precedence matrix* for a graph with n vertices is an $n \times n$ matrix $[m_{ij}]$ such that:

m_{ij} = a mark if and only if there exists an arc in the graph to vertex x_i from x_j, or $i = j_j =$ blank otherwise

A *mark* is any symbol other than a blank. (The marks on the diagonal are always circled.)

The rows and columns of this matrix correspond to the vertices of the graph and thus to the tasks. The marks correspond to arcs. Thus we will often talk about marks and arcs interchangeably.

Figure 3.4a shows a graph and its corresponding precedence matrix.

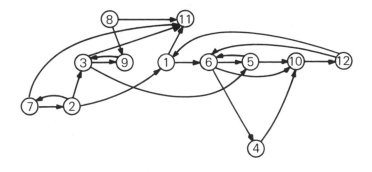

	1	2	3	4	5	6	7	8	9	10	11	12
1	⊗	X										X
2		⊗					X					
3		X	⊗						X			
4				⊗		X						
5			X		⊗	X						
6	X				X	⊗						X
7		X					⊗					
8								⊗				
9			X					X	⊗			
10				X	X	X				⊗		
11	X		X				X	X			⊗	
12										X		⊗

Figure 3.4a

Illustration of Procedure 3.2 Partitioning. *Step 1:* I = 0. *Step 2:* Vertex 8 has no predecessor, which we recognize in the matrix by noting that row 8 has no off-diagonal mark. *Step 4:* We assign the order number I to vertex 8 and remove it and its arcs. We cross out row and column 8 in the matrix. The row number I is written above the circle for vertex 9. The results of this process are shown in Fig. 3.4b.

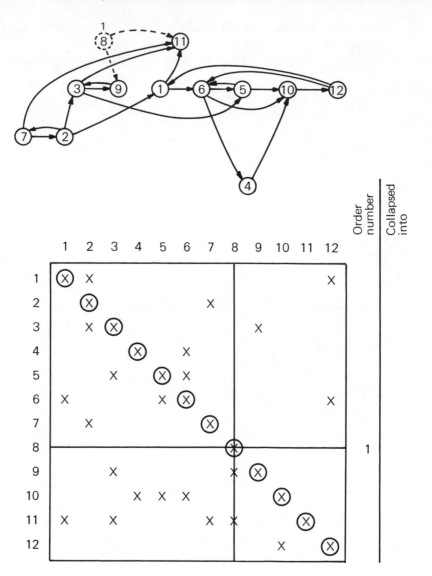

Figure 3.4b

Illustration of Procedure 3.2 Partitioning. *Step 2:* Every row has an off-diagonal mark. *Step 3:* We arbitrarily begin with row (vertex) 1 to generate a sequence of predecessors. By arbitrarily choosing the lowest numbered vertex as the next predecessor we generate the sequence 1,2,7,2. This reveals the circuit (2,7,2). Thus we collapse 7 into 2. We use a + mark to show the marks introduced into row and column 2 by collapsing, then cross out row and column 7. The results of this process are shown in Fig. 3.4c.

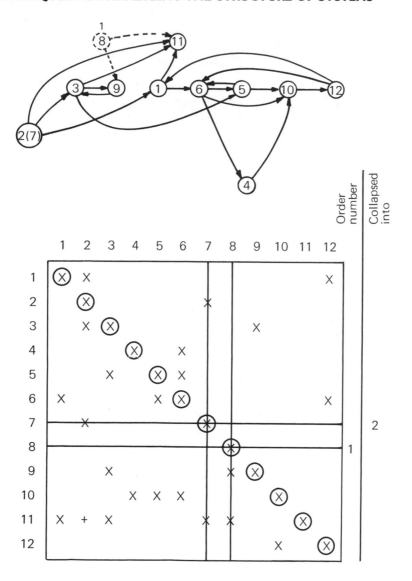

Figure 3.4c
Illustration of Procedure 3.2 Partitioning. *Step 2:* Now row 2 has no off-diagonal mark.
Step 4: We assign row 2 and the row which collapsed into it, i.e., row 7, the next order
numbers, 2 and 3. We remove vertex 2 and its arcs by crossing out row and column 2 to
obtain Fig. 3.4d.

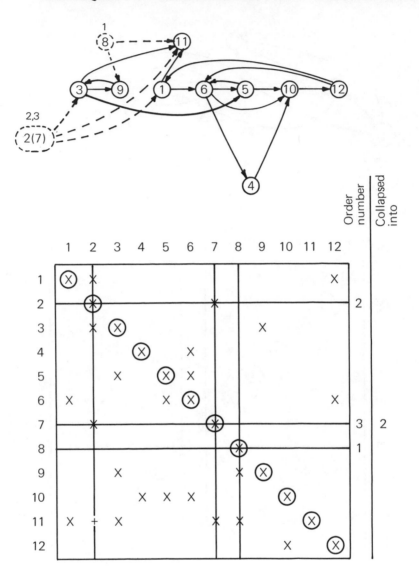

Figure 3.4d

Illustration of Procedure 3.2 Partitioning. *Step 2:* Every row has an off-diagonal mark. *Step 3:* We generate the sequence of predecessors 1,12,10,4,6,1. This reveals the circuit (1,6,4,10,12,1). We collapse 4,6,10 and 12 into 1 to obtain Fig. 3.4e.

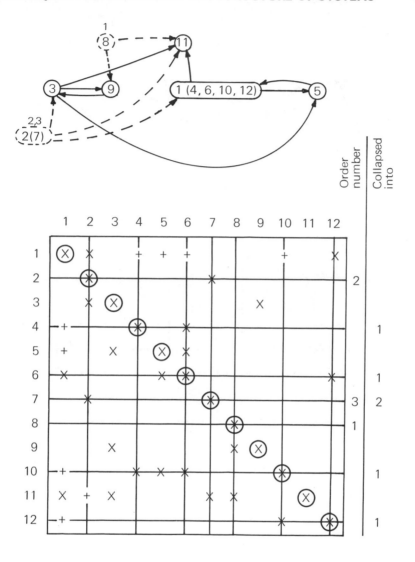

Figure 3.4e

Illustration of Procedure 3.2 Partitioning. *Step 2:* Every row has an off-diagonal mark. *Step 3:* We generate the sequence of predecessors 1,5,1 and thus collapse 5 into 1 to obtain Fig. 3.4f.

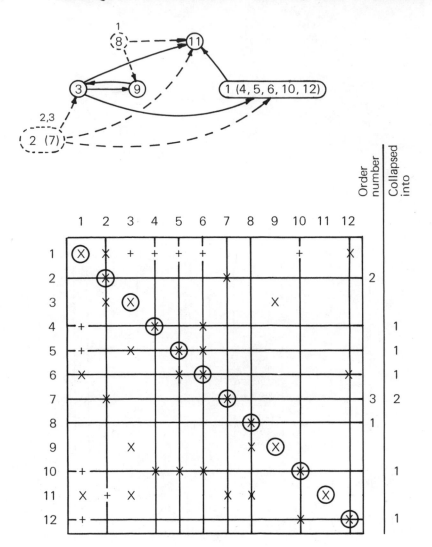

Figure 3.4f

Illustration of Procedure 3.2 Partitioning. *Step 2:* Every row has an off-diagonal mark. *Step 3:* We generate the sequence of predecessors 1,3,9,3 and collapse 9 into 3 to obtain Fig. 3.4g.

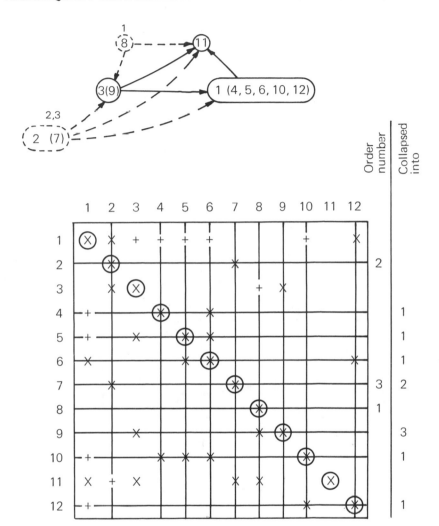

Figure 3.4g

Illustration of Procedure 3.2 Partitioning. *Step 2:* We see that row 3 has no off-diagonal mark. *Step 4:* We assign to row 3 the order number 4, and row 9 which collapsed into 2 is assigned the order number 5. Cross out row and column 3. *Step 2:* Then row 1 has no off-diagonal mark. *Step 4:* We assign to row 1 the order number 6, and assign to the rows which collapsed into it, i.e., 4,5,6,10 and 12, the order numbers 7,8,9,10 and 11, respectively. We cross out row and column 1. *Step 2:* Finally row 11 has no off-diagonal mark. *Step 4:* We assign to row 11 the order number 12 and cross out row and column 11. This gives Fig. 3.4h.

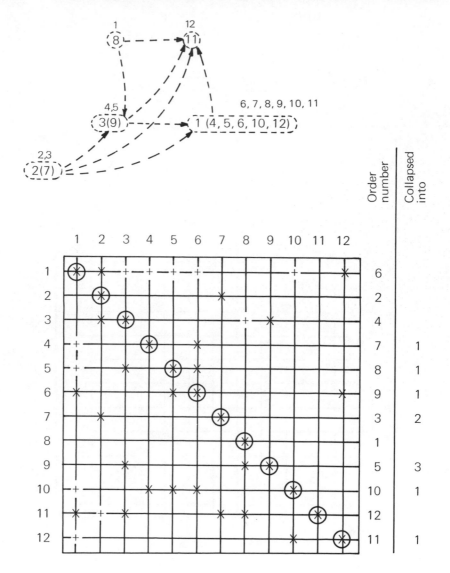

Figure 3.4h

Illustration of Procedure 3.2 Partitioning. *Step 2:* No vertex remains. *Step 5:* An order number has been assigned to each vertex. In Fig. 3.4i we will reorder the rows and columns according to these order numbers to obtain the block triangular form of the matrix.

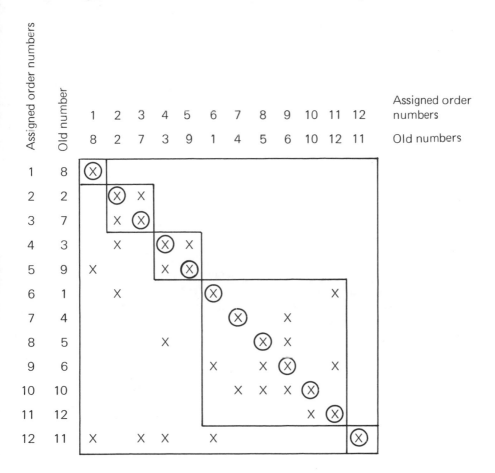

Figure 3.4i
Block Triangular Form of Matrix. From this matrix it can be seen that (1) tasks within a block are assigned contiguous integers. (2) No arc goes from a high-numbered task in one block to a low-numbered task in another block, i.e., there is no mark above the blocks on the diagonal. The third item of the definition cannot be observed so readily. However, the reader can trace paths to satisfy himself that (3) given any two tasks in the same block there is a path in each direction between them.

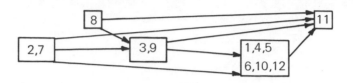

Figure 3.4j
Graph of the Partition of the Graph in Fig. 3.4a

Marks occur on the diagonal since $x_i \le x_i$ for all x_i. We circle the diagonal marks to make the diagonal easy to distinguish. Off-diagonal marks in row i represent immediate predecessors to x_i, i.e., arcs leading into x_i, and off-diagonal marks in column i represent immediate successors, i.e., arcs leading out of x_i.

To use the matrix to trace a path beginning with vertex x_j, we find an off-diagonal mark in column j. Assume this mark occurs in row i. This says there is an arc in the corresponding graph from x_j to x_i. Find column i by moving across row i to the diagonal. Now go down column i, etc. This process on the matrix corresponds to tracing paths on the graph.

Below is a procedure for partitioning. We first describe the operation of the procedure in terms of a graph, and later describe the equivalent operations on a precedence matrix. In Figure 3.4 an example of the procedure is shown using both the graph and the precedence matrix. In the procedure the terms *predecessor* and *immediate predecessor* are used. x_i is a predecessor of x_j if there is a path from x_i to x_j, and is an immediate predecessor if there is an arc directly from x_i to x_j.

PROCEDURE 3.2.   *Partitioning* (see Figure 3.4)

Step 1: Set $I = 0$.

Step 2: If no vertex remains, go to Step 5. If every remaining vertex has an immediate predecessor, go to Step 3. Otherwise, go to Step 4.

Step 3: (Find a Circuit) Choose any remaining vertex. Choose an immediate predecessor vertex, then an immediate predecessor of this vertex in turn, etc. Continue this process until some vertex has been encountered twice. This must occur since by assumption every vertex has an immediate predecessor and one cannot continue indefinitely to encounter new vertices if the number of vertices in the graph is finite. By this process a path has been traversed, backwards, from the repeated vertex back to itself. This path is a circuit. Choose one vertex in the circuit and collapse all the vertices of the circuit into that one vertex. By "collapsing" is meant the following: One vertex of the circuit, chosen arbitrarily, will be made to *represent* the set of all the vertices in the circuit. This representative vertex will now have an arc to or from a vertex in the remaining graph if and only if some vertex in the circuit had an arc to or from, respectively, the other vertex. All the vertices and their arcs in the circuit other than this representative vertex are removed from the graph. Then we say these vertices have been "collapsed" into the representative vertex. Go to Step 2.

Step 4: (Number and Remove a Vertex) Choose a vertex with no immediate predecessor. Remove this vertex and all the arcs which enter or exit from it. If this removed vertex represents n original vertices (see Step 3), number these n original vertices $I + 1$ to $I + n$. Let $I = I + n$. Go to Step 2.

Step 5: This completes the procedure. The blocks are the sets of vertices which are removed in Step 4. Step 4 assigns contiguous numbers to the vertices in a block. Since a vertex (or block of vertices) is not numbered until all predecessors have been removed and assigned lower numbers, no arc goes from a high-numbered vertex in one block to a low-numbered vertex in another block. Since vertices are collapsed into a block only when they are in a circuit with another vertex already in the block, a path can be found from any vertex in the block to any other vertex in the block by following subpaths of these circuits.

If there are no circuits, this procedure reduces to Procedure 3.1.

Although it is easier to understand the partitioning procedure in terms of the graph, a matrix is more compact and easier to work with. In the matrix a vertex with no predecessor appears as a row with no off-diagonal mark. Removing a vertex in the graph corresponds to crossing out the row and the corresponding column in the matrix. We write the number I assigned to this vertex in the column labeled "Order Number." Vertex x_k would be collapsed into vertex x_j as follows: add a mark in row x_j in every column that a mark occurred in row x_k; add a mark in column x_j in every row that a mark occurred in column x_k; cross out row and column x_k. Next to row x_k we write a j in the column labeled "collapsed into" to indicate x_k was collapsed into x_j.

If the rows and columns of the precedence matrix are reordered according to the numbering assigned in Procedure 3.2, then the new matrix is block triangular with square blocks on the diagonal (see Figure 3.4i). These blocks contain sets of tasks which correspond to the blocks of the partition of the graph. For a given ordering of the tasks within the blocks, the marks above the diagonal in the block correspond to feedback arcs for this ordering. In practice we will work with precedence matrices which represent the graphs rather than with the graphs themselves.

DEFINITION.   Given a graph $G = (X,U)$, the *graph of the partition of G* is the graph $G_p = (X_p,U_p)$ where X_p is the set of blocks $[x_i]$ which partition X, and an arc $([x_i],[x_j]) \in U_p$ if and only if there exist $x_a \in [x_i]$ and $x_b \in [x_j]$ such that $(x_a,x_b) \in U$. ($[x_i]$ denotes the set of vertices equivalent to x_i.)

If in Step 4 of the above procedure we did not actually remove the vertex, but just removed it from consideration as a predecessor to another vertex, then we would obtain the graph of the partition (see Figure 3.4j). We will refer again to the graph of a partition in a theorem in chapter 5.

Now, with this ordering, all feedback arcs are between tasks in the same block. In the next process, tearing, we consider one block at a time. The set of contiguous integers assigned to the tasks within each block are reordered to affect which arcs are the feedback arcs. This is considered in the next section.

3.2 Tearing

To reorder the tasks within a block we can proceed in either of two basic ways: (1) We can assign an ordering to the tasks, then determine for this ordering which arcs are feedbacks by seeing which arcs go from high-numbered vertices to low-numbered vertices. Then we evaluate these feedback arcs using our knowledge of the semantics they represent. We can then assign another ordering, etc., until we obtain an ordering with a set of feedbacks which we find to be acceptable when considering their semantic meaning. (2) Or we could remove (tear) arcs where our knowledge of the semantics indicates the arcs represent acceptable feedbacks, and where our knowledge of the structure indicates that the removal of the arc would be effective in breaking circuits. Procedure 3.2 is used to test whether any circuit remains. If any circuit remains, there will be a block of size greater than one. Then we tear more arcs or tear another set of arcs and test again. We do this until no circuit remains and we have an ordering for the tasks. These removed arcs will be called *tear arcs* or *tears*. These tears, or a subset, correspond to feedbacks for the ordering obtained.

We use primarily the second of these two approaches. In general, it is not easy to determine a priori precisely what sets of torn arcs will leave a graph with no circuit. However, chapter 4 examines the use of so-called shunt diagrams to show which arcs when torn have the greatest effect on breaking circuits.

DEFINITION.   We remove a set of arcs and partition the remaining tasks by Procedure 3.2 to determine whether sufficient arcs have been torn so that no circuit remains. This process is called *tearing*. The arcs removed are said to be "torn." If the removal of a set of arcs which are feedback arcs for some ordering leaves the remaining graph with no circuit, then this set of arcs is called a *complete feedback tear set*.

A procedure is now presented to test the tearing to see if it breaks all the circuits, and, if so, assigns numbers to all the vertices. How one might select the tears will be discussed in chapter 4. Steps 2, 3, and 4 test whether these tear arcs form a complete feedback tear set for some ordering and, if so, obtains such an ordering. Step 5 concerns how to proceed if this test is not met.

PROCEDURE 3.3.   *Tearing* (removing arcs and testing whether a circuit remains)

Given a block, proceed as follows:

Step 1: Select a set of tears. (Chapter 4 discusses the use of shunt diagrams as an aid in making this selection.)

Step 2: Remove the tear arcs from the graph.

Step 3: Partition this remaining graph by Procedure 3.2. Any tear arc (mark) in the original graph (precedence matrix) from a task now in a high-numbered block to a low-numbered block is called a *feedback tear arc (mark)*. The set of feedback tear arcs is a subset of the set of tear arcs.

Step 4: If there are no blocks in this partition of the remaining graph which contain more than one task, then the set of tear arcs selected in Step 1 contains a subset which is the complete feedback tear set for this ordering and we are finished. If there are blocks in this partition of the remaining graph which contain more than one task, go to Step 5.

Step 5: We can now proceed in either of the following ways:

a.   We consider each block in this partition which contains more than one task and proceed to tear this block just as we did with the original block, starting with Step 1 above. The feedback tear arcs obtained in this process are added to the set of feedback tear arcs obtained for the original block. We continue this process until we exit in Step 4.

b.   We return to Step 1 to make a new selection of tear arcs. Chapter 4 will show that we can now add stronger constraints to the making of this selection through the use of so-called generalized shunts.

The shunt diagram (developed in chapter 4) is printed by the computer for each block. This shunt diagram will communicate to the user what combinations of arcs satisfy certain necessary conditions to be a complete feedback tear set for some ordering. The user usually seeks the smallest set or some other criteria which satisfies these and other semantic conditions he wishes to impose. The challenge in the design of the computer program is to make the necessary conditions implied by the shunt diagrams as strong as possible without making the diagrams too complicated to use. By using a multiple-step process of "tear and try" it is possible to get by with weaker and less complex necessary conditions in order to make the system easier to use.

The chosen tear marks are given numbers (*level numbers*) by the user to communicate to the computer program how the tearing is to occur. The computer program treats these level numbers as follows:

PROCEDURE 3.4.   *Partitioning Using Level Numbers for Marks to Show Tears*

First the whole matrix is partitioned, treating all of the marks equally. Then, block by block of the partition, the highest numbered remaining marks are removed and the block is partitioned again. This process of removing the highest numbered remaining marks and partitioning each remaining block is continued until a partition is completed with all the remaining marks having the same, lowest level number. Finally, the removed marks are restored to their original row and column in this new ordering. This has the effect of producing blocks within blocks where the marks with the smaller level numbers are confined to the smaller blocks. (See Figure 2.7.)

There are two bases used to assign level numbers to marks in the matrix. We refer to these as a priori and a posteriori assignment.

For the a priori assignment we consider the semantic meaning of the tasks and precedence constraints. We assign numbers in a high range (e.g., from 6 to 9) to those marks we would accept above the diagonal. Within this range we assign the highest numbers to those marks which would be more tolerable to tear and lower numbers to those marks which would be less tolerable to tear. We consider the semantics to make these decisions. A mark above the diagonal indicates where an assumption or estimate must be used for the predecessor. Thus, one would generally consider it more tolerable to tear a mark, and thus assign it a high-level number, if the likely error in the assumption or estimate of the predecessor represented by the mark would not have a significant effect on the task it precedes. This may occur because a good estimate can be made, or because the task is insensitive to the error in the estimate. Further discussion of the criteria for a priori assignment is deferred to the chapters on applications.

Using these a priori assignments of level numbers, we then use Procedure 3.4. If, after partitioning with all the remaining marks at the same level, we still have one or more blocks of size greater than one, then we proceed to a posteriori assignment of level numbers. Here we make a tearing analysis using shunt diagrams to gain insight into the structure of these remaining blocks. We use this insight to choose tears which would be effective for removing circuits. We evaluate the possible choices with our knowledge of the technical meaning of the marks. By this means we assign level numbers in a low range (e.g., from 1 to 5). (This will be discussed further in chapter 4.)

To the extent that a designer of models or systems can assign level numbers a priori which order the preferences (or reluctances) for the marks to be torn, he may be able to reduce the size of the blocks to be torn a posteriori by using shunt diagrams. On the other hand, it is sometimes not easy to review and assign a priori numbers to all marks in a large system. It may be preferable to defer these judgments until one has first obtained the partition and shunt diagram so that the designer's attention is confined to those marks which the shunt diagram suggests as good candidates for feedback tears.

We do not pretend to obtain optimal solutions to the ordering problem. The evaluation of tear sets may be based upon the semantic evaluation of combinations of feedback arcs which can only be done by the user. Thus there may be no computer-representable a priori criteria by which to define an optimum. Furthermore, it is possible that our techniques may fail to expose tear sets which the user would evaluate on a semantic basis as superior to those exposed by the technique. However, to paraphrase Saaty's (1959) definition of operations research, we wish to find satisfactory solutions where otherwise only worse solutions are economically available.

As Herbert Simon (1969) points out, a salesman does not stay at home because he knows that there is no satisfactory algorithm by which he can afford to obtain an optimal solution to the traveling salesman problem. The cost of obtaining an optimum solution might be more than he could save. Furthermore, an optimal solution to an idealized traveling salesman problem would probably not satisfy various practical constraints imposed on his trip.

Thus, he uses the best heuristic procedures available to him to choose a solution, and, if he can afford it, makes the trip. (Krolak et al. have developed heuristic approaches to the traveling salesman problem from this point of view [Krolak, Felts and Marble:71].) A satisfactory solution is a solution one can afford to obtain and can afford to use.

3.3 Transitive Closure

Another mathematical construct that needs to be developed for later use is transitivity. The rule of transitivity states: If there is a path from j to k and a path from k to i, then there is a path from j to i. If we add to the original matrix a mark in every position i,j where a path from j to i is implied by transitivity, we have the transitive closure of the original matrix. It is often useful to obtain this transitive closure; for example, if the marks in the matrix represent causes and effects, the transitive closure shows all the resulting effects (see chapter 9).

Given a matrix, how do we generate this transitive closure? The easiest way is to use the following operations. If row i has a mark in column k, then i is preceded by k and thus by everything that precedes k. So we put a mark in row i in every column where k has a mark but row i doesn't. If this operation is done for every row in the matrix and is repeated until no more changes occur, we will have produced the transitive closure. If the matrix is lower triangular, this operation need be done only once on each row.

We are used to ordering the matrix so it becomes block triangular (demonstrated in Figure 3.5). How does this affect our procedure? We note that every element within any block can be reached from every other element within that block. Thus, transitive closure will fill in every position within the block. We have not filled these positions in Figure 3.5 because we already know the elements within a block are all interdependent. To fill the block would obscure its structure.

If there is a mark to the left of the block, then transitivity implies a mark be added in that column to every row of the block. This is shown by the b's in Figure 3.5b. For example, the mark in 12,6 produces the b's in column 6 rows 8,2, and 5. Similarly, if there is a mark below the block, transitivity implies a mark be added in that row in every column of the block. For example, the mark in 3,12 produces the b's in row 3 columns 8, 2, and 5.

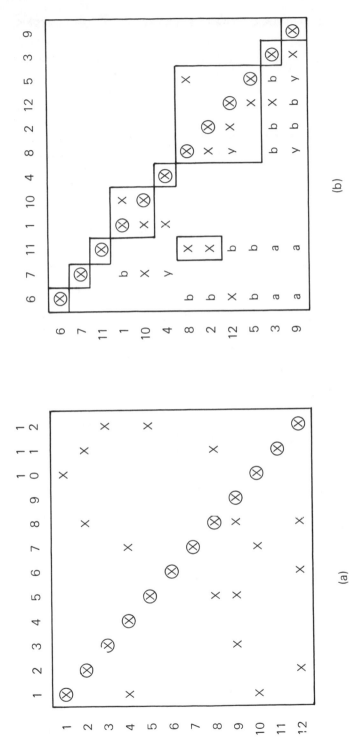

a = marks added by transitivity

b = marks added by transitivity
of block

y = redundant marks

☒
☒   all but one mark redundant
by transitivity of block

(a)

(b)

Figure 3.5
(a) Original matrix. (b) Matrix reordered with transitive closure

If transitivity produces a new mark where one already existed, the mark is replaced by a *y*. This implies that the original matrix showed a path of length one as well as one or more longer paths implied by transitivity. It is sometimes important to note when we have these "redundant" paths (see chapter 9).

A box is drawn around the marks in column 11 rows 8 and 2 to indicate that either of these marks would imply the other by the transitivity due to the block.

Looking at the last row in Figure 3.5b, we can see that not every vertex is a predecessor of the vertex in the last row. Thus there is more than one connected graph. We can see from the last row of the matrix that there is a connected graph including 6, 11, 8, 2, 12, 5, 3, and 9. Looking at the last row that does not include these elements, we find the other connected graph which includes 7, 1, 10, and 4. This is confirmed by looking at Figure 3.6.

3.4 A More Careful Look at Earlier Examples

Let us look more closely at examples of two classes of applications we looked at briefly in chapter 2: the analysis of the procedures for the engineering design of systems, and the analysis of the design of econometric models. The first involves information flows and precedence relations, while the second concerns systems of simultaneous nonlinear equations. These diverse applications should illustrate the range of applicability of these techniques and suggest other possible applications to the reader.

ANALYSIS OF THE ENGINEERING DESIGN OF SYSTEMS

In this consideration of the procedures for the engineering design of systems (studied in more detail in chapter 7), the tasks are the determination of design variables, the execution of computer codes and design procedures, or the preparation of documents. It is characteristic of engineering design that the precedence relations among the tasks contain circuits, e.g., variable *A* cannot be determined until variable *B* is known or assumed, but variable *B* cannot be determined until variable *A* is known or assumed. Such a circuit is usually handled by assuming either *A* or *B* to initiate an iteration through successive designs. It is the primary role of

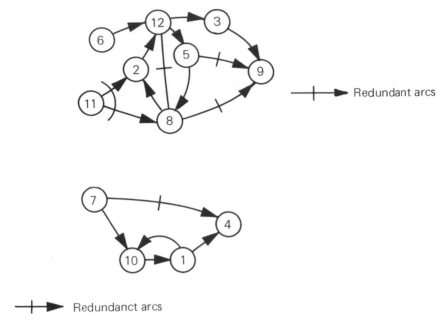

Redundant arcs

Redundanct arcs

Figure 3.6
Graph of Fig. 3.5 Showing Two Connected Graphs and Redundant Arcs

engineering to resolve these circuits, usually by iterating through trial designs with pencil and paper or the computer, so as to minimize or eliminate the more costly iteration which might otherwise occur during fabrication.

The feedback arcs are arcs from tasks which are assumed in order to start the design process, to other tasks which require them. The methods of partitioning and tearing are used to plan the iterative design procedure.

Look back at Figure 2.4, which is a precedence matrix representing the design of an electric car. A mark, either 0 or 9 in this case, in row i column j implies that in order to determine variable x_i, variable x_j must first be known or assumed. Circled x's mark the diagonal. The 9's represent our a priori choice of tears.

Figure 2.5 shows the block triangular matrix obtained by reordering rows and columns according to Procedure 3.2. The marks above the diagonal are confined to a square block on the diagonal. This is the smallest block which will include all the circuits involving these variables.

Originally, when Figure 2.4 was constructed, the 9's were used to indicate those marks which represent where the user feels adequate estimates can be made for the results of tasks before they are completed. 0's are used for the remaining marks in the matrix. (The higher level number marks are torn, then successively lower level number marks are torn, etc. as in Procedure 3.4.) Since these marks are assigned a high number, they are chosen first for tearing. Removing the 9 marks, partitioning the remaining block, reordering the rows and columns according to this partition, then restoring the 9's to their proper row and column positions produces Figure 2.6.

In Figure 2.6, after some further analysis using shunt diagrams, level number 5 has been assigned to two marks chosen a posteriori using shunt diagrams (see chapter 4). Then by Procedure 3.4 the variables within this block have been reordered (see Figure 2.7). Now we see that only the 5's and 9's remain above the diagonal. One of these 5's is confined to a block of size 6, while the other is further confined to a block of size 2.

The matrix as ordered in Figure 2.7 now represents a satisfactory design procedure. We begin a preliminary design by making estimates for variables where they are indicated by marks above the diagonal. The column shows the variable estimated. The row shows where it is used. When one gets to the last variable in a block, a design review is required to determine whether the estimates were valid, or whether new estimates should be used to make another design iteration.

Note that when any design variable changes, by looking down columns one can quickly determine what other variables must be redetermined as a consequence so that plans and schedules can be made for the resulting work.

ANALYSIS OF ECONOMETRIC MODELS

The second class of applications we consider deals with systems of equations and is applied here to the formulation, analysis, and diagnosis of econometric models (considered in more detail in chapter 8).

We apply partitioning and tearing to econometric models composed of n equations which may be solved for the values of n unknown variables. (Chapter 8 also considers input-output models.) The econometrician calls these unknown variables the current endogenous variables. Past values of these variables are called lagged endogenous variables. Variables which are given, and thus not solved for in the system, are called exogenous. At each point in time the n equations are solved for the n current endogenous variables, given known values for the lagged endogenous and the exogenous variables.

Figure 2.8 shows a matrix where rows represent equations and columns represent current endogenous variables. The marks show where current endogenous variables appear in the n equations. This particular example is the revised version of the Klein-Goldberger (1955) annual model of the United States economy.

Note that unlike Figure 2.4, the rows and columns are drawn from different sets. As we will see in chapter 5 this matrix represents not a directed graph, but a bipartite graph which maps the set of variables onto the set of equations. Let us circle one mark in each row so that one and only one circle occurs in each column. This process corresponds to assigning a unique dependent variable to each equation. Now we permute the rows so that the circles are brought to the diagonal. This process gives Figure 2.9. We label each row by the variable assigned to it. We now have a mapping of the set of variables onto itself. This gives a precedence matrix for a directed graph which can be partitioned and torn as in the design problem above. In Figure 2.10 the variables have been reordered to obtain a square block of size 14 on the diagonal.

An off-diagonal mark represents the occurrence of an independent variable in an equation. For a given ordering, a mark above the diagonal represents a feedback. It may sometimes be desirable while diagnosing the behavior of the model to simplify the model by removing these feedbacks. This can be done by replacing the occurrence of a variable where it appears as a feedback by an estimate obtained from outside the model. One might, for example, use for estimates actual historical values for these variables, or a time series, a constant, or some other simple model to generate values of the variable. This device simplifies the model by removing the dependence of the equation on a later equation.

We seek to find tears such that: (1) the behavior of the model is not significantly affected by a replacement of an occurrence of a variable with an estimate obtained from outside the system, and (2) the remaining matrix can be reordered to obtain a block triangular matrix such that the blocks on the diagonal represent meaningful submodels. This allows the individual submodel to be studied without the complications due to circuits between submodels.

It will be shown later that the sets of variables within the blocks of Figure 2.10 are unique and do not depend upon the assignment of the dependent variables to the equations. However, when we come to analyze the equations within each of the blocks to obtain feedback tear sets, we will see that this process is not independent of how we have assigned the dependent variables to the equations. Therefore, this becomes the subject of some consideration in chapter 5.

Within the block in Figure 2.10 level numbers have been assigned to several marks to be torn. We chose these tears by considering how reasonable it would be to approximate these variables as they appear in these equations by exogenous variables, i.e., estimates obtained from outside the system, and thereby eliminate the circuits. The remaining x's are considered to be level 0. In Figure 2.11 the variables have been reordered by Procedure 3.4 to reveal a reasonable structuring of the model into submodels. If Y is assumed to be a known function of time so that the equations which use Y do not depend on equation 7, the model partitions into submodels as follows: a submodel involving four variables (A_1, p, P_c, S_p), a submodel involving two variables (D, K), and fourteen submodels of one variable each.

4 Techniques for Tearing Systems

4.1 Generation and Application of Shunt Diagrams

This chapter discusses shunt diagrams which will help reveal various possible combinations of tears which satisfy certain necessary conditions for eliminating all circuits. Users can then apply their technical insight to choose from among these tear combinations. Procedure 4.1 defines shunt diagrams by showing how they are generated, and Procedure 4.3 shows how they are used for the selection of tears.

First we make two observations:

OBSERVATION 4.1.   A complete feedback tear set must include at least one arc in each circuit.

OBSERVATION 4.2.   A lower bound on the number of tears in any complete feedback tear set is the number of circuits which can be found such that no two circuits have an arc in common.

(Woodmansee [1968] has conjectured that for only a rather special class of graphs is $K(G) < J(G)$ where $K(G)$ is the maximum number of arc disjoint circuits and $J(G)$ is the minimum number of tears in a complete feedback tear set. Thus it appears that $K(G) = J(G)$ for an extensive class of graphs.)

Motivated by these observations, we consider one block at a time and proceed as follows:

PROCEDURE 4.1. *Generate a set of shunt diagrams for a block.*

Follow these steps in Figure 4.1. Consider just the vertices and arcs within one block.

Step 1: (Find a Principal Circuit) Choose a vertex. Choose an immediate predecessor of this vertex, then an immediate predecessor of this vertex in turn, etc., until some vertex has been encountered twice. (This is the same as Procedure 1.2, Step 3.) Reversing the sequence of vertices from the repeated vertex back to itself gives a circuit. Such a circuit is called a *principal circuit*. Remove the arcs in this circuit.

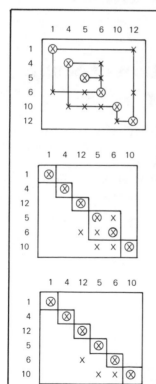

Alternate a:

Step 1: We can begin, arbitrarily, with any vertex. We would ordinarily begin with vertex (row) 1. But to lead us into conditions which will illustrate certain points we will begin with row 5. There we find a predecessor mark in column 6. We go to row 6 to find a predecessor mark in column 1, etc. Thus we generate the sequence of predecessors 5, 6, 1, 12, 10, 4, 6. We note that vertex 6 is repeated. This gives a principal circuit (6,4,10,12,1,6). (Later we will show an alternate shunt diagram we could get if we were to make different arbitrary choices in tracing the predecessors.)

Step 1 (continued): We remove the marks in the above matrix which correspond to the arcs in this principal circuit (6,4,10,12,1,6). We then partition the remaining matrix to obtain the matrix to the left. In the block remaining in this matrix we generate the sequence of predecessors 5, 6, 5. This gives a second principal circuit (5,6,5).

Step 1 (continued): The matrix to the left contains just those marks which remain after removing the marks in the two principal circuits. No block of size greater than one remains after partitioning this matrix. Thus there are no more principal circuits to be found.

Step 2: We use this last matrix to generate shunts for each of the principal circuits. Consider the principal circuit (1,6,4,10,12,1), which is the same as the above principal circuit but with the smallest vertex first. We seek paths between pairs of vertices in this circuit which do not include any other vertex in the circuit. Consider the vertices in the order they occur in this principal circuit. We trace successors, i.e., go down columns instead of across rows. Going down column 6 we see it has an arc leading to 10. 10 is in the principal circuit. So we have the shunt path (6,10). Neither 4 nor 10 has an arc leading out of it. 12 has an arc to 6, which is in the principal circuit, giving the shunt (12,6). 1 has no arc leading from it. We ignore the arc (5,10) because it does not go between vertices of the same principal circuit. For each of these shunts the length of the path in the principal circuit is 2, while the length of the path is 1. Thus the index of both shunts is 1. The shunt diagram for each principal circuit is shown next to Step 3.

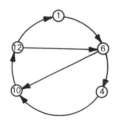

Step 3: The principal circuit (1,6,4,10,12,1) is written in sequence down a column. The shunt (6,10) is written in a column with a *B* opposite 6 in the principal circuit, and an *E* opposite the 10. In this case there is no other vertex in this shunt. An *I* is written to fill the space between *B* and *E*. Similarly, shunt (12,6) is written in the next column, the *B* opposite 12 and the *E* opposite 6, with the *I* to fill the position between.

This graph corresponds to the shunt diagram for the principal circuit (1,6,4,10,12,1).

A shunt diagram can also be written for the principal circuit (5,6,5). There is no shunt. This is so trivial that ordinarily no shunt diagram would be drawn.

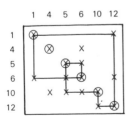

Alternate b:
Generation of First Principal Circuit and Resulting Shunt Diagram

Alternate b, Step 1: Here we start with 1 and generate the sequence of predecessors 1, 12, 10, 5, 6, 1. This is different from the sequence generated in the first example. This gives the principal circuit (1,6,5,10,12,1). (In this case it happened that the first vertex was repeated. This is not always so.) This example gives us the opportunity to illustrate several other situations which did not occur in the first example.

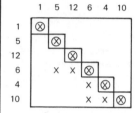

Alternate b, Step 2: We remove the marks in the above matrix which correspond to the arcs in the principal circuit (1,6,5,10,12,1). We partition the remaining matrix and find that no block of size greater than one remains. Thus there is only one principal circuit this time. Following Procedure 4.2, we will use this matrix to generate shunts for this principal circuit.

(See the illustration of Procedure 4.2 which shows the generation of shunts used in Alternate b, Step 3 below.)

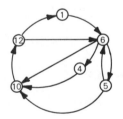

	Shunts			
	0	1	1	3 Indices
1			1	1
6	B	B	E	E
5	4	1		B
10	E	E		1
12			B	1

Alternate b, Step 3: The principal circuit (1,6,5,10,12,1) is written in sequence down a column. The shunt (6,4,10) is written with the *B* opposite the 6 in the principal circuit and the *E* opposite the 10. The 4 is written in the space (fence) from the *B* to the *E*. The shunts (6,10), (12,6), and (5,6) are written in successive columns.

This graph corresponds to the shunt diagram for the principal circuit (1,6,5,10,12,1).

Figure 4.1
Illustration of Procedure 4.1
Generate a Set of Shunt Diagrams for a Block

Partition the remaining graph of this block. If there remains any block with more than one vertex, repeat Step 1 for each such block. Otherwise, go to Step 2.

Step 2: For each principal circuit, consider each pair of distinct vertices in the principal circuit. Using just those arcs remaining in the graph of this block after Step 1, generate all paths between pairs of these vertices in this principal circuit which do not include any other vertex in the principal circuit. Such a path is called a *shunt*. [Procedure 4.2 shows how to do this.] The length of the path in the principal circuit between these vertices minus the length of the shunt is called the *index* of the shunt. If a shunt has a negative index, then interchange the shunt with the path in the principal circuit between these same two vertices to obtain a longer principal circuit. Then start again in Step 2 using this new principal circuit. Thus each time we obtain a negative index we lengthen the principal circuit until the index of all shunts is nonnegative.

Step 3: For each principal circuit, beginning arbitrarily with some vertex, print the vertices in the principal circuit in sequence down a column. Print each shunt in a new column with a *B* in the same row as the vertex in the principal circuit where the shunt begins, and an *E* in the same row as the vertex in the principal circuit where the shunt ends. (When tracing either the principal circuit or a shunt, we always proceed downward cyclically, i.e., going off the bottom and continuing at the top if necessary.) The interval from the *B* to the *E* is called the *fence*, while the interval from the *E* to the *B* is called the *gap*. In the fence, print the sequence of vertices between the beginning and the end vertices of the shunt, if any. Fill the remaining positions in the fence with *I*'s. Sort the shunts so they appear in successive columns in increasing order of their indices. This is a *shunt diagram*.

Consider the 6 × 6 block remaining after the process of partitioning in Figure 3.4i.

Procedure 4.2 generates all possible shunt paths beginning with a given vertex. A path is followed until it can go no further. Then one backs up to the next previous vertex in the path until another branch is found that was not taken before. This branch is followed in the same way. When one finally backs up to the original vertex and there are no other branches that can be taken, the process is finished.

PROCEDURE 4.2. *Generate shunt paths* (see Figure 4.2)

Use the matrix of the block which remains after all arcs in the principal circuits have been removed.

For each vertex in the principal circuit, remove all checks, go to the column corresponding to this vertex, then begin with Step 1.

Step 1: Find an unchecked mark in this column. If there is no such mark, go to Step 3. If this mark is in a column corresponding to a vertex in the principal circuit, go to Step 4. Otherwise, go to Step 2.

Step 2: Put a check next to this mark and add the vertex corresponding to this row to the path being generated. Move along this row to the diagonal, i.e., move to the column corresponding to this row. Go to Step 1.

Step 3: Erase the last vertex in the path being generated and go to the column corresponding to what is now the last vertex in the path. If no vertex remains in the path, go to Step 5. Otherwise, go to Step 1.

Step 4: Check this mark and add the vertex corresponding to this row to the path. Record this path as a shunt. Go to Step 3.

Step 5: All possible paths beginning with this vertex in the principal circuit have been generated until they either come to a dead end or intersect the principal circuit to produce a shunt. Proceed to the next vertex in the principal circuit and begin again with Step 1.

By construction, the principal circuits are a set of arcwise disjoint circuits. Therefore, the number of principal circuits is a lower bound on the number of tears in a complete feedback tear set.

DEFINITION. *A principal graph* is that subgraph whose vertices and arcs are those of a principal circuit and its shunts.

DEFINITION: Given two vertices x_i and x_j and two paths $(x_i,x_j)_1$ and $(x_i,x_j)_2$ with no vertices in common except x_i and x_j, then any subpath of $(x_i,x_j)_1$ and any subpath of $(x_i,x_j)_2$ are said to be *parallel*. Given paths $(x_j,x_i)_1$ and $(x_i,x_j)_2$ with no vertices in common except x_i and x_j, then any subpath of (x_j,x_i) and any subpath of (x_i,x_j) are said to be

We shall illustrate how Procedure 4.2 is used to find shunt paths. Look at the matrix with Alternate b, Step 2 of Procedure 4.1. Let us begin the illustration tracing shunt paths from vertex 6. (Note that we are tracing shunts moving toward successors—along columns.)

Step 1: Going down column 6 we find that an unchecked off diagonal mark in row 4. 4 is not in the principal circuit.

Step 2: We check this mark (row 4, column 6). We write down the path (6,4). We go to column 4.

Step 1: In column 4 we find an unchecked mark in row 10. 10 is in the principal circuit.

Step 4: Check this mark (row 10, column 4). The path (6,4,10) is recorded as a shunt. The length of the path in the principal circuit between 6 and 10 is 2. The length of the shunt is also 2. Therefore the index of the shunt is 0.

Step 3: Erasing the last vertex in the path leaves the path (6,4). Then go to column 4.

Step 1: There is no unchecked mark in this column.

Step 3: We erase the 4 in the path, leaving just (6), and go to column 6.

Step 1: There is an unchecked off diagonal mark in this column at row 10; 10 is in the principal circuit.

Step 4: Check this mark in row 10, column 6. We add 10 to the path to obtain (6,10) and record this path as a shunt.

Step 3: Erase vertex 10 from the path, leaving just (6). Go to column 6.

Step 1: There is no unchecked mark in this column.

Step 3: When we erase this vertex there is now no vertex left in the path.

Step 5: We have now traced all possible shunt paths from vertex 6 in the principal circuit.

Similarly, starting from vertices 12 and 5 in the principal circuit, we generate the shunts (12,6) and (5,6).

We have now used Procedure 4.2 to generate the shunts as required in Step 2 of Procedure 4.1.

Figure 4.2
Illustration of Procedure 4.2
Generate Shunt Paths

antiparallel. (See the last graph in Figure 4.1. Arc (4,10) is parallel to (6,5) as both are subpaths of paths from 6 to 10. Arc (4,10) is antiparallel to (12,1) because (4,10) is a subpath of a path from 6 to 10 and (12,1) is a subpath of the path from 10 to 6.)

OBSERVATION 4.3.   Consider a principal graph and a set of tears in this graph. Necessary conditions for these tears to eliminate all circuits in this graph (and thus to include a complete feedback tear set) are:

1.   There is at least one tear in the principal circuit. Otherwise the principal circuit remains.
2.   There is no one untorn shunt parallel to all tears in the principal circuit. Otherwise this untorn shunt and its antiparallel path in the principal circuit form a circuit.

OBSERVATION 4.4.   Consider each principal graph. Since each arc, and thus each circuit in this graph, appears in the original graph, then a necessary condition to tear all circuits in the original graph is to tear all circuits in each such principal graph.

Our procedure is to find a set of tears for each principal circuit and its shunts which meet the conditions of Observation 4.3. By Observation 4.4 we take the union of these sets of tears as the set of tears we use in Step 1 of Procedure 3.3. Then we test the remaining graph by Procedure 3.3 to determine whether any circuits remain.

OBSERVATION 4.5.   The extension of a horizontal line drawn on the shunt diagram between two adjacent vertices of a given arc in the principal circuit will "hit," i.e., intersect the fence of, each shunt and only those shunts parallel to the given arc in the principal circuit. We say this tear in the principal circuit lines up with the fence in the shunt. Similarly, this line will hit the gap of each shunt antiparallel to the given arc in the principal circuit. We say then that the tear in the principal circuit lines up with the gap in the shunt.

OBSERVATION 4.6.   Combining observations 4.3 and 4.5, we have the following more easily applied necessary conditions to eliminate all circuits in the principal graph:

1.   There must be at least one tear in the principal circuit.
2.   The following conditions must be met for each shunt in the diagram. Either (a) there is a tear in the principal circuit such that the tear lines up with the gap in this shunt, or (b) some arc in this shunt is torn.

We proceed as outlined below to choose our tears using the shunt diagram so that the necessary conditions of Observation 4.3 hold.

PROCEDURE 4.3.   *Using shunt diagrams to choose tears which satisfy the necessary conditions of observation 4.3* (see Figure 4.3.)

Consider a shunt diagram for a principal circuit and its shunts.

Step 1: Select one or more arcs to be torn in the principal circuit. For each arc in the principal circuit being torn, draw a horizontal line through the principal circuit between the vertices of the arc being torn. Carry these lines across the diagram. Mark each shunt which is hit by all of these lines. If there is no such marked shunt, go to Step 3. Otherwise, go to Step 2.

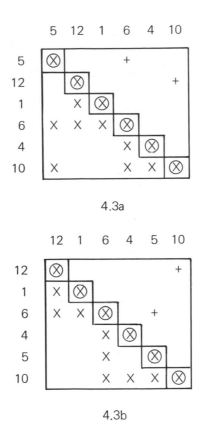

4.3a

4.3b

Figure 4.3
Illustration of Procedure 4.3

Step 2: Consider each of these marked shunts. Each is parallel to all of the torn arcs in the principal circuit. We can now proceed in either of the following ways: (a) "tear the shunt," i.e., tear an arc in the shunt, then remove the mark from this shunt, or (b) tear an arc in the principal circuit which lines up with the gap in this shunt. Remove the mark from this shunt and any other marked shunt whose gap lines up with this new tear in the principal circuit.

Step 3: This completes the procedure. We now have a set of tears for this principal graph, which satisfy the conditions of Observation 4.3.

This procedure is repeated for each principal graph. The union of the tears from these principal graphs then is used in Step 1 of Procedure 3.3.

Consider the shunt diagram resulting from Alternate b of Figure 4.1.

Principal Circuit	Shunts (Step 2)			
	0	1 1	3	Indices
1		1 1		
6	B B	E E		
5	4 1	B		
10	E E	1		(Step 1)
12		B 1		

Principal Circuit	Shunts (Step 4)			
	0	1 1	3	Indices
1		1 1		
6	B B	E E		(Step 3)
5	4 1	B		
10	E E	1		
12		B 1		

Step 1: Let us tear the arc (10,12) in the principal circuit. A horizontal line is drawn across the diagram between vertices 10 and 12 in the principal circuit.

Step 2: This line intersects the fence of only one shunt, (5,6). Thus this shunt is parallel to the tear in the principal circuit. We check this column. (Any other tear in the principal circuit would have been parallel to two or more shunts.)

Step 3: We choose another tear in the principal circuit (6,5), which is not parallel to the marked shunt.

Step 4: We remove the mark above the shunt (5,6) since this shunt is not parallel to the new tear in the principal circuit. Now no shunt is marked, so no shunt is parallel to all the tears in the principal circuit. Thus the conditions of Observation 4.3 are satisfied. This gives the tears (10,12) and (6,5).

We apply Procedure 3.3 to determine whether any circuit remains. Procedure 3.3 partitions this matrix with the tear marks removed. No blocks of size greater than one remain after this partition, confirming that this is a complete tear set. We have indicated where the tear marks were with + signs.

Below we show an alternate choice for the second tear which could have been used in Step 3 above.

Step 3: (alternate 1)
Instead of choosing a second tear in the principal circuit as above we have chosen to tear the marked shunt.

Principal Circuit	Shunts				(Step 4)
	0	1	1	3	Indices
1			I	I	
6	B	B	E	E	
5	4	I		B	(Step 3)

10	E	E		I	

12			B	I	

Step 4: (alternate 1)

We remove the marks above the shunt which we just tore. Now there is no untorn shunt which is parallel to all the tears in the principal circuit. Thus we have satisfied the observations of Condition 4.3. This gives the tears (10,12) and (5,6).

We apply Procedure 3.3 to partition the matrix with these tears removed to determine whether they constitute a complete feedback tear set. Since no blocks of size greater than one remain, we indeed do have a complete feedback tear set.

Now we have two possible complete feedback tear sets. There are of course many more possibilities. These two look like reasonably good possibilities since they have only two tears. One would have to apply his understanding of the technology of the problem to determine which of these two tear sets is the better, or whether neither is any good and he should seek another tear set even though it may have more tears.

Using Shunt Diagrams to Choose Tears
Which Satisfy the Necessary Conditions of
Observation 4.3.

4.2 Criteria for Choosing Tears

There are several criteria which the user may apply in the selection of tears, subject to overriding considerations based upon his knowledge of the semantics of the tasks and constraints. These criteria, their motivation, and the heuristic techniques which contribute to achieving them are as follows:

1. *Criterion:* A small number of tears.

    Motivation: If each tear represents where an approximation or initial guess is to be used, reducing the number of tears reduces the number of errors introduced.

    Heuristic: Tear an arc in the principal circuit which is parallel to the least number of shunts. Mark the shunts, if any, which are parallel to all tears in the principal circuit. Tear another arc in the principal circuit which is parallel to the least number of marked shunts, or else tear a parallel shunt if only one remains. Sometimes it is possible to tear an arc which we observe to be common to several shunts. See Figure 4.4.

2. *Criterion:* Tend to confine the most tears to the smallest square boxes on the diagonal.

    Motivation: If there are to be iterations within iterations, these inner

Principal Circuit	0	1	1	3
1			I	I
6	B	B	E	E
5	4	I		B
10	E	E		I
12			B	I

	5	12	1	6	4	10
5	⊗			+		
12		⊗				+
1		X	⊗			
6	X	X	X	⊗		
4				X	⊗	
10	X			X	X	⊗

Figure 4.4
Small Number of Tears

iterations are done more often. It is desirable to confine the inner iterations to a small number of tasks; then fewer tasks are involved in each inner iteration.

Heuristic: Tear arcs in the principal circuit such that the lowest index parallel shunt, if any, has as high an index as possible. (Note: if the block is of size N and the lowest index shunt marked has an index I, then we might expect that the size of the block which would remain after this tearing would be $N - I$. (However, this is not always the case because of what will be called *overlooked circuits* in section 4.3.) Partition, then tear the remaining blocks; see Figure 4.5.

Principal Circuit	0	1	1	3
1			I	I
6	B	B	E	E
5	4	I		B
10	E	E		I
12			B	I

	12	1	6	5	4	10
12	⊗					+
1	X	⊗				
6	X	X	⊗	+		
5			X	⊗		
4			X		⊗	
10			X	X	X	⊗

Figure 4.5
Confine Most Tears to Small Blocks

Principal circuit	0	1	1	3		4	5	10	12	1	6
1			I	I	4	⊗					+
6	B	B	E	E	5		⊗				+
5	4	I		B	10	X	X	⊗			+
10	E	E		I	12			X	⊗		
12			B	I	1				X	⊗	
					6		X		X	X	⊗

Figure 4.6
Tears in a Small Number of Different Columns

3. *Criterion:* The tears should appear in a small number of different columns.

   Motivation: A minimum number of different variables are approximated or estimated to begin the iteration. Each of these approximations or estimates may be used as a predecessor to more than one task.

   Heuristic: If we find several shunts with a *B* opposite the same vertex in the principal circuit, then by tearing all the arcs which exit from this vertex we have torn the principal circuit and each of these shunts; see Figure 4.6.

Principal circuit	0	1	1	3		6	4	5	10	12	1
1			I	I	6	⊗	+			+	+
6	B	B	E	E	4	X	⊗				
5	4	I		B	5	X		⊗			
10	E	E		I	10	X	X	X	⊗		
12			B	I	12				X	⊗	
					1					X	⊗

Figure 4.7
Tears in a Small Number of Different Rows

4. *Criterion:* The tears should appear in a small number of different rows.

Motivation: Estimates or approximations directly affect the least number of different tasks.

Heuristic: If we find several shunts with an *E* opposite the same vertex in the principal circuit, then by tearing all the arcs which enter this vertex we have torn the principal circuit and each of these shunts; see Figure 4.7.

4.3 Strengthening the Necessary Conditions

We may choose tears according to Procedure 4.3, but find when we partition that one or more circuits still remain. These circuits are called *overlooked circuits*. This occurs because Procedure 4.3 finds tears that

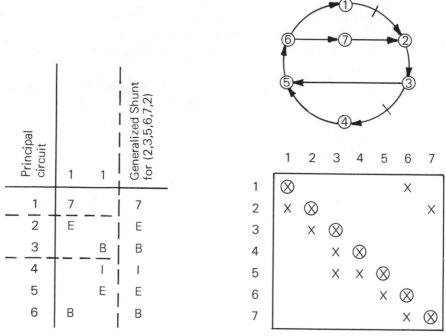

Figure 4.8
Example of the Insufficiency of Observation 4.3. The set of tears (1,2) and (3,4) satisfy the conditions of Observation 4.3. However, circuit (2,3,5,6,7,2) remains. The addition of the generalized shunt formed from this circuit would cause one to tear either (2,3) or (5,6). Note that the generalized shunt can be formed by putting together in one column shunts which have no common parallel arc in the principal circuit.

satisfy necessary conditions to tear all circuits, but these conditions may not be sufficient. Figures 4.8 through 4.10 show examples of tears that satisfy the necessary conditions but still leave circuits untorn.

When this occurs we can find the overlooked circuits and use them to make what we call generalized shunts. With these generalized shunts added to the shunt diagram, any new choice of tears will be forced to break these overlooked circuits.

The arcs in a shunt, plus the arcs in the principal circuit corresponding to the gap in the shunt, form a circuit. Thus an overlooked circuit is represented by a shunt with gaps corresponding to arcs in the circuit that occur in the principal circuit. Since the overlooked circuit may contain more than one sequence of arcs in the principal circuit, the generalized shunt may have more than one gap; see figures 4.8 and 4.9.

In Figure 4.10 we see that there are two principal circuits, each with a generalized shunt containing arcs from the other. In addition, these shunts have negative indices. In such a case it is often possible to combine the two principal circuits into one; see Figure 4.11.

So, in summary, we choose a tear according to Procedure 4.3. If the partitioning shows there still exist one or more circuits, we use them to

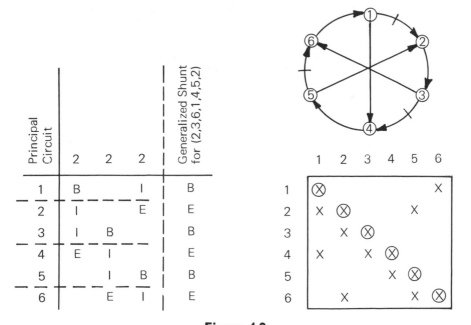

Figure 4.9
Another Example of the Insufficiency of Observation 4.3. The set of tears (1,2), (3,4) and (5,6) satisfy the conditions of Observation 4.3. However, the circuit (2,3,6,1,4,5,2) still remains. Adding the generalized shunt for this circuit would lead one to tear either (2,3), (4,5) or (6,1) in the principal circuit. (This is an example of a graph for which K(G) < J(G).)

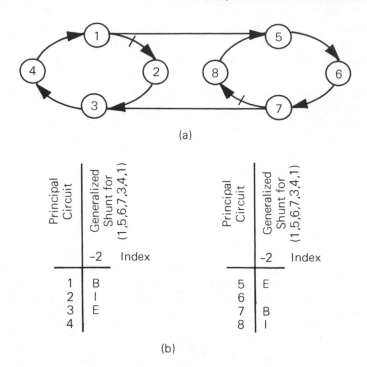

(a)

(b)

Figure 4.10

Arcs between Principal Circuits. (a) Consider the two principal circuits (1,2,3,4,1) and (5,6,7,8,5). Tears (1,2) and (7,8) would satisfy Observations 4.3 and 4.4, but the circuit (1,5,6,7,3,4,1), which contains arcs from both principal circuits, would remain. (b) The circuit (1,5,6,7,3,4,1) is represented by a generalized shunt for each principal circuit. Tearing either principal circuit in the gap of this shunt (between the E and the B) tears this circuit. Note: If we were to tear (2,3) in the first principal circuit this would have the effect of tearing the corresponding generalized shunt of the second principal circuit.

make generalized shunts which we add to the shunt diagram and find another set of tears. This process may be continued until we obtain a set of tears which leaves no circuit.

   Usually it is not worth the effort required to find all overlooked circuits and add their corresponding generalized shunts to the original shunt diagram. Generally, we proceed as in Procedure 3.3 for tearing, beginning with a shunt diagram having no generalized shunts. If we arrive at Step 5—namely, we are left after partitioning with one or more blocks containing more than one task—then we make a shunt diagram for these remaining blocks and either (a) use the shunt diagram of these remaining blocks to generate more tears which we then add to the original set of tears, or (b) add generalized shunts to the original shunt diagram to represent all

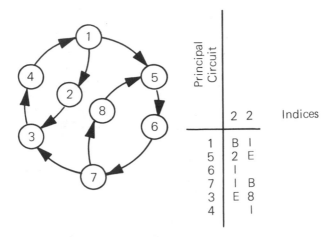

Figure 4.11
Two Principal Circuits Transformed into One. The indices of the generalized shunts in
Fig. 4.10 are negative. This suggests that a larger principal circuit can be found, in this case by
combining both principal circuits into one. This larger principal circuit and its shunt diagram
are drawn to the left. However, it is not always possible to avoid the problems of connected
multiple principal circuits by this device.

the overlooked circuits which now show up as principal circuits and shunts
in the shunt diagrams for these remaining blocks. With these added
constraints we return to the original shunt diagram to make another choice
of tears. These two alternatives correspond to alternatives a and b in
Procedure 3.3.

Thus, the shunt diagram gives the user significant insight into the
effect of tearing various combinations of arcs. Conceivably, stronger
conditions than those implied by the shunt diagram might be developed.
However, we do not see how this could be done without making the system
more expensive or more difficult to use. Thus, we have chosen the
successive application of a "tear and try" procedure using weaker and less
complex necessary conditions to avoid a system which might otherwise be
too complex to be practical.

Examples of the applications of shunt diagrams using the output of
the TERABL program are presented in the last sections of chapters 7, 8,
and 9.

5 The Analysis of Systems of Equations

5.1 Output Assignment

This chapter extends the concept of partitioning to cover systems of n equations in n variables by associating a precedence matrix with the system of equations. The precedence matrix is obtained by assigning an output variable in each equation.*

DEFINITION. The *structural matrix* S for a system of n equations in n unknowns is an $n \times n$ matrix $[s_{ij}]$ where:

s_{ij} = a mark if variable v_j appears in equation e_i
= blank otherwise

Note that $[s_{ij}]$ is an *incidence matrix* which shows the mapping of the set of variables onto the set of equations. This corresponds to what is called a *bipartite graph* with arcs from vertices in one set representing the variables to vertices in another set representing the equations.

Note that when speaking of the structure of a system of equations we consider which variables occur in which equations, but not how they appear. The structural matrix may represent either a linear or a nonlinear system of equations.

*This chapter develops more carefully some of the material introduced in Steward (1962).

DEFINITION. An *output assignment* is the choice of one mark in each row of a structural matrix $[s_{ij}]$ such that one and only one mark is chosen in each column. The chosen marks are circled. The term "output" refers to the variable assigned to an equation, i.e., the choice of dependent variable. (Note that the output assignment is a one-to-one mapping of the set of equations onto the set of variables.)

Figure 5.1 shows a system of equations, its structural matrix, and its corresponding bipartite graph.

Figure 5.2a shows the structural matrix of Figure 5.1b with an output assignment marked by circled x's. Next to each row we have placed the letter designating the column which is assigned as the output of the equation in that row. Above each column is the number of the row to which this column has been assigned as output. If we permute the rows so that the order of the variables labeling the rows is the same as the order of the variables labeling the columns, as in Figure 5.2b, we have a precedence

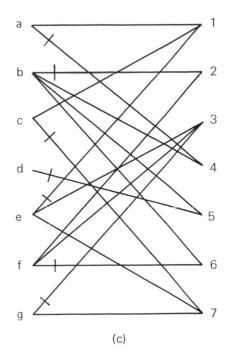

$$F_1(x_a, x_c, x_e) = 0$$
$$F_2(x_b, x_f) = 0$$
$$F_3(x_e, x_f, x_g) = 0$$
$$F_4(x_a, x_b) = 0$$
$$F_5(x_b, x_d) = 0$$
$$F_6(x_b, x_f) = 0$$
$$F_7(x_c, x_e, x_g) = 0$$

(a)

	a	b	c	d	e	f	g
1	X		X		X		
2		X				X	
3					X	X	X
4	X	X					
5		X		X			
6		X				X	
7			X		X		X

(b)

Figure 5.1
(a) Systems of equations. (b) Structural matrix. (c) Bipartite graph (outputs marked with slash).

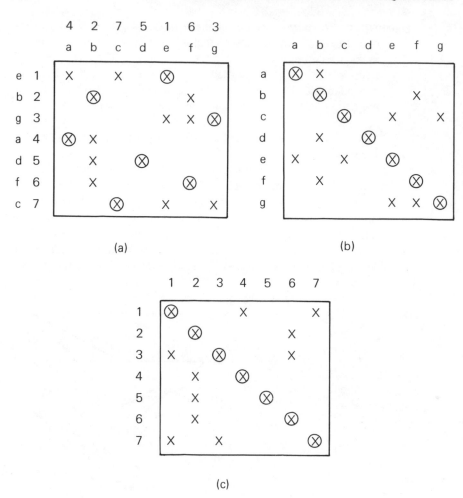

Figure 5.2

(a) Structural matrix with outputs circled. (b) Variable precedence matrix. (c) Equation precedence matrix.

matrix showing the mapping of the set of variables onto itself. (Note that the effect is to permute the rows so as to bring the assigned outputs to the diagonal.) Also, the columns could be permuted to bring the outputs to the diagonal. This produces the precedence matrix showing the mapping of the set of equations onto itself (Figure 5.2c).

DEFINITION.   Given a structural matrix and an output assignment, if the rows (equations) are permuted to bring the outputs to the diagonal, we have a *variable precedence matrix* showing the mapping of the set of variables onto itself. If the columns (variables) are permuted to bring the outputs onto the diagonal, we have an *equation precedence matrix* showing the mapping of the set of equations onto itself. (We will most often use the variable precedence matrix.)

From either of these precedence matrices we can draw a graph, as in Figure 5.3. Note that the two graphs are the same except for the labeling of the vertices. Here we show both labelings.

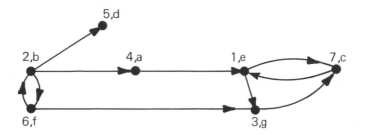

Figure 5.3
Graph of System of Equations with Output Assignment le, 2b, 3g, 4a, 5d, 6f, 7c
Showing Both Variable and Equation Labeling

Now that we have a precedence matrix and its corresponding graph, it can be partitioned as shown in chapter 3. The graph of the partition for the graph of Figure 5.3 is shown in Figure 5.4.

Using this graph of the partition we can put the precedence matrix in block triangular form as in Figure 5.5. Note that we have labeled rows and columns by both equation labels and their corresponding variable labels. If we assume the equation labels for the rows and the variable labels for the columns, we have the original structural matrix with both the rows and columns permuted to put it in block triangular form with the output marks occurring on the diagonal.

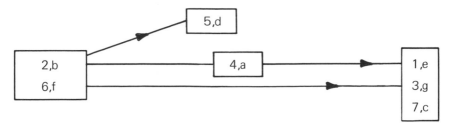

Figure 5.4
Graph of Partition of System of Equations

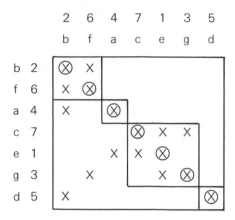

Figure 5.5
Block Triangular Form

5.2 The Uniqueness of the Partition

Now, given a particular output assignment, we have shown a partition and
the graph of the partition for a system of equations. However, as we can
easily see, an output assignment need not be unique. For example, we may
substitute 2f and 6b for 2b and 6f, respectively, in the output assignment in
Figure 5.2a to obtain a new output assignment. In order for the partition
and the graph of the partition to be a property of the system of equations
alone, it must be independent of the choice of output assignment. We shall
now develop some of the machinery which will be used to prove that
indeed this is the case.

DEFINITION.   Consider the structural matrix with an output assignment
and a series of connected lines drawn on the matrix as follows: A
horizontal line drawn from a mark which is not an output mark to the
output mark in the same row will be called a *horizontal link*. A
horizontal link drawn from column i (representing variable v_i) along
row j (representing equation e_j) to the output variable assigned to that
row (represented by v_j) is written $[v_i \, e_j \, v_j]$. A vertical line drawn in
column j from an output mark to another mark in the same column
(say, this other mark is in row k) will be called a *vertical link* and is
written $[e_j \, v_j \, e_k]$, where again v_j is the variable assigned as output to
equation e_j. A sequence of connected alternating horizontal and
vertical links with common equations between successive links is a
chain. A chain can begin or end with either a vertical or horizontal

link. The redundant variable and equation symbols are dropped; that is, the link $[v_i \, e_j \, v_j]$ followed by $[e_j \, v_j \, e_k]$ is written as the chain; $[v_i \, e_j \, v_j \, e_k]$. (Each vertical link followed by a horizontal link in a chain corresponds to an arc in the precedence matrix.) A chain which begins and ends on the same mark is a *loop*. (A loop corresponds to a circuit in the precedence matrix.) Figure 5.6a shows a chain and 5.6b shows a loop. (Note that we draw lines along rows or columns from one mark to another mark, but ignore the other marks in the row or column which the line happens to pass through.)

Now we can use the concepts of links, chains, and loops to show the relationship between one output assignment and another.

Consider Figure 5.7a. Let us try to move the output in column c from row 7 to row 1. We indicate this move by writing the link [7c1]. Now there are two outputs in row 1. We draw the line represented by the horizontal link [c1e] to assign another output in this row. Then we move the output e from 1 to 3, represented by the vertical link [1e3], which leaves two outputs in row 3. We move the output in row 3 from e to g, [e3g], and the output in column g from 3 to 7, [3g7]. Since row 7 was vacated by our original move, we again have just one output assigned to each row and just one output assigned to each column. We call this process of changing from one output assignment to another *output relabeling;* see Figure 5.7b. Here the output relabeling occurred around the loop [7c1e3g7]. Clearly, the output

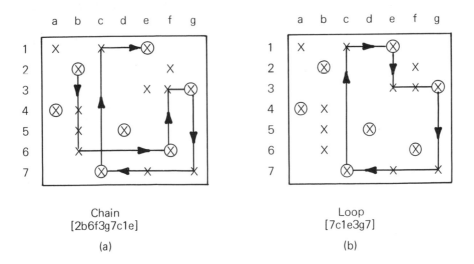

Chain
[2b6f3g7c1e]

(a)

Loop
[7c1e3g7]

(b)

Figure 5.6

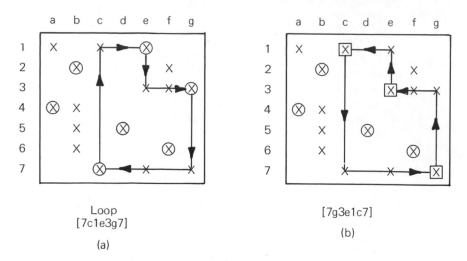

Loop
[7c1e3g7]

(a)

[7g3e1c7]

(b)

Figure 5.7
(b) After relabeling

relabeling must move along a chain until it returns to the row vacated by the original move. This leads to the following observation:

OBSERVATION 5.1.   Any transformation from one output assignment to another must be accomplished by relabeling of outputs around one or more loops. Note also that relabeling around a loop causes the direction of that loop to reverse. For example, the loop [7c1e3g7] becomes the loop [7g3e1c7] after the relabeling.

DEFINITION.   A loop used in the output relabeling is called a *relabeling loop* and a circuit corresponding to this loop is called a *relabeling circuit*.

Given a structural matrix and an output assignment, we have seen that there is a corresponding precedence matrix and its graph. A chain from one variable to another (or from one equation to another) on the structural matrix with a given output assignment corresponds to a path between these two variables (or equations) on the precedence matrix and its graph. Consider the chain [2b6f3g7c1e] in Figure 5.8a. Dropping the equations from the chain, which are implied by the output assignment anyway, we have the path (b,f,g,c,e). We can then follow this path in the variable precedence matrix in Figure 5.8b. This relation between the chains and loops in the structural matrix and the paths and circuits in the precedence matrix, respectively, is central to the theorem below.

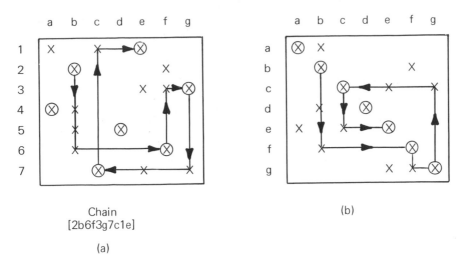

Chain
[2b6f3g7c1e]

(a)

(b)

Figure 5.8
(a) Same as Fig. 5.6a. (b) Variable precedence matrix showing path (b,f,g,c,e) corresponding to chain in part (a).

THEOREM 5.1.   *The partition and the graph of the partition of the precedence matrix derived from assigning outputs to a system of* n *equations in* n *unknowns is independent of the output assignment.*

Proof: Relabeling the output assignment, because it occurs around a loop, does not affect the existence of a chain from j to i in the structural matrix although a chain, if it does exist, may go around the opposite side of one or more loops. Thus the relabeling does not affect the existence of a path from j to i in the precedence matrix. Then, $x_j \leq x_i$ after the change of output assignment if and only if $x_j \leq x_i$ before the change. The partition based upon $x_j \leq x_i$ then remains the same.

We now consider the following theorem on the existence of output assignments:

THEOREM 5.2.   *If there does not exist an output assignment for a system of equations with as many variables as equations, then there exists a subset of equations with fewer variables than equations.**

*This theorem has been proven in the context of distinct representatives of sets by P. Hall (1935) and by M. Hall (1956). The proof appears in Steward (1962). It is repeated here because the techniques are germane to some procedures developed later in this chapter.

Proof: The proof is constructive and takes the form of the flowchart in Figure 5.9. The reader should follow the flowchart as one would follow the steps of a theorem. The flow exits in one of two conditions—either an output assignment has been made (T_1), or else there is a subset of equations which contains fewer variables than equations (T_2).

To complete the proof it must be shown that the process cannot fall into an infinite circuit. To do this we apply the ideas developed in this book to tear the flowchart. If we tear the lines leaving box 12, no circuit will remain. This can be seen by noting that except for these torn lines no box has a line entering it from a higher numbered box. Now we establish that the flow can go through box 12 at most n^2 times where n is the number of rows (or columns) in the structural matrix.

Checks on the columns can only be removed in box 10. We can go through box 10 at most n times because in order to go through box 10 we must also go through box 8. But every time we go through box 8 a new equation is assigned, which can be done at most n times without removing any marks. Every time we leave box 12 we check another column, which can be done at most n times for each time the marks are removed in box 10. Therefore, the process represented in this figure must exit after a number of steps proportional to n^2 in one of the two conditions T_1 or T_2.

Note that the nature of the algorithm presented in this theorem is such that when a step is reached where more than one output is assigned in the same column, a relabeling occurs along a chain which ends in an unassigned column. We shall make further use of this concept later in the chapter.

As an aside we note the following theorem related to linear systems:

THEOREM 5.3.   *A linear system of* n *equations in* n *unknowns for which there does not exist an output assignment has no unique solution.*

Proof: By definition the value of the determinant is the sum of terms with appropriate sign where each term is composed of the product of elements such that there occurs exactly one element from each row and exactly one element from each column. But the choice of elements in such a term corresponds to an output assignment. If there exists no output assignment, then such a non-zero term does not exist. Thus the determinant is zero, implying that the system of linear equations does not possess a unique solution.

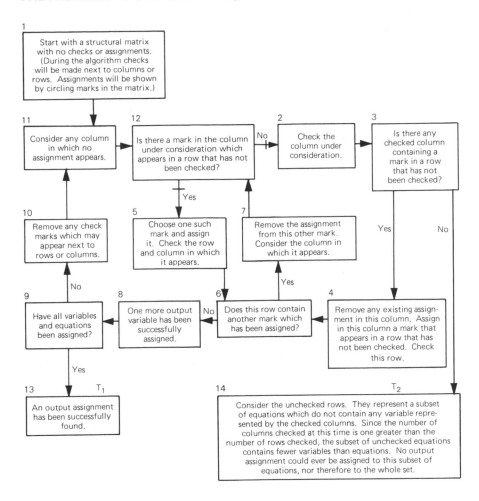

Figure 5.9
An Algorithm which Finds an Output Assignment (T₁) or Else Shows that None Exists. Thus, also, a constructive proof for the conditions under which an output assignment exists.

Although, in general, the existence of an output assignment is neither necessary nor sufficient for the existence or uniqueness of a solution of a system of nonlinear equations, as shown below, it is a useful guide for focusing on the solvability of systems of equations. Assume we are given a system of n equations in n variables which does not possess an output assignment. Then, as we have shown, it possesses a subset of equations with fewer variables than equations. (This subset is called "overdetermined" in Weil and Steward:67.) This usually means there are

too many equations to constrain these variables, leading to the possibility
that the extra constraints may not be consistent. Consider the variables
which do not appear in this subset of equations. The number of these
variables is greater than the number of equations they appear in. This
usually means there are insufficient equations to constrain these remaining
variables [Marschak:50]. (This complementary subset of equations is
called "underdetermined" by Weil and Steward [1967]. They call the
system itself "structurally indeterminate.")

   These conditions, however, refer only to the structure of the system
of equations, not to how the variables appear in the equations. When we
consider how the variables appear in the equations it becomes possible for
a structurally indeterminate system to possess a unique answer. Consider
the following system of equations:

$$x^2 = -y^2$$
$$z = 1$$
$$2z = 2$$

The second and third equations form a subset which is overdetermined.
The first equation is an underdetermined subset. If x and y are allowed to
be complex, we can see that no unique solution exists. However, if we add
the constraint that x and y must be real, then we see that the only possible
solution in real space of the first equation is $x = y = 0$. The second and
third equations are overdetermined but consistent. Thus this set of
equations does possess a unique solution $x = y = 0$, $z = 1$ even though
they are structurally indeterminate.

   In most applications the conditions required for a structurally
indeterminate system to possess a unique solution would be unusual. It is
most often useful when considering whether a system of equations
possesses a unique solution to establish first whether it has an output
assignment. If it does not, then one should look carefully at the
overdetermined subset to see if the extra constraints are consistent, and
look at the underdetermined subset to see whether some condition on the
solution space may offer the necessary added constraints. Otherwise, the
system of equations most likely does not possess a unique solution.

5.3 Relation of the Structure to Means of Solving Simultaneous Equations

Now we can relate this theory to the actual process of solving systems of n equations in n unknowns.*

The system of equations in Figure 5.10 can be solved as follows: Equation 1 is a function of one variable. If it can be solved at all, it can be solved independently of the other equations. Thus we solve equation 1 immediately for the number represented by x_a. Then this number is substituted into equation 2. Equation 2 may then be solved for the number represented by x_b, which is substituted into equation 3 which can then be solved for the number represented by x_c. Figure 5.11 shows this sequence of operations.

$$F_1\,(x_a)\ \ = 0 \qquad 1$$
$$F_2\,(x_a,x_b) = 0 \qquad 2$$
$$F_3\,(x_b,x_c) = 0 \qquad 3$$

	a	b	c
1	X		
2	X	X	
3		X	X

Figure 5.10
System of Equations and Their Structural Matrix (No Loop)

This sequence of substitutions may be represented by a chain. Substitution is represented by a vertical link, and solution of an equation for a dependent (output) variable given an independent variable is represented by a horizontal link. The sequence of substitutions for the solution of the system of equations in Figure 5.10 is represented by the chain [1a2b3c]. This is shown on the structural matrix in Figure 5.12.

$$F_1(x_a)\ \ = 0 \Rightarrow G_{1a} = x_a$$
$$\qquad\qquad\qquad\qquad\downarrow \text{number}$$
$$F_2(x_a,x_b) = 0 \Rightarrow \quad G_{2b}(x_a) = x_b$$
$$\qquad\qquad\qquad\qquad\qquad\downarrow \text{number}$$
$$F_3(x_b,x_c) = 0 \Rightarrow \quad G_{3c}(x_b) = x_c$$

Figure 5.11
Solution of system of equations by numerical solution

*The remainder of this chapter has also appeared in Steware:62.

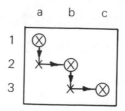

Figure 5.12
The Chain [la2b3c] Representing a Sequence of Substitutions

However, consider the equations in Figure 5.13. We assume the output assignment 1a, 2b, 3c. Here we do not have an equation involving only one variable that we can solve first; in fact, the chain representing the sequence of substitutions forms a loop. A different output assignment would change the direction but not the existence of the loop. The existence of any loop implies that the equations in the loop must be solved simultaneously. Clearly, simultaneity implies an equivalence relation, e.g., if A and B must be solved simultaneously, and B and C must be solved simultaneously, then A, B, and C must be solved simultaneously. We can see that this simultaneous set corresponds to a block in the partition of the set of equations.

A simultaneous set of equations in a loop (a block) can be solved by either of two methods: iteration or elimination. If we consider iteration as the substitution of numbers, and elimination as the substitution of symbolic expressions, we will see below that, given the same output assignment, both methods imply the same sequence of substitutions. These substitutions may be represented by chains.

Solving the equations in Figure 5.13a using the output assignment 1a, 2b, 3c we have Figure 5.14.

$$F_1\,(x_a,x_c) = 0 \qquad 1$$

$$F_2\,(x_a,x_b) = 0 \qquad 2$$

$$F_3\,(x_b,x_c) = 0 \qquad 3$$

Loop [1a2b3c1]

Figure 5.13
System of Simultaneous Equations Containing a Loop

$$G_{1a}(x_c) = x_a$$
$$G_{2b}(x_a) = x_b$$
$$G_{3c}(x_b) = x_c$$

Figure 5.14
The equations given the output assignment 1a, 2b, 3c

To solve this system by iteration we begin by assuming a value of x_c^o and proceed to iterate as in Figure 5.15.

$$x_c^o \text{ (estimate)}$$

$$G_{1a}(x_c^i) = x_a^i$$

$\downarrow$ number

Number $G_{2b}(x_a^i) = x_b^i$

$\downarrow$ number

$$G_{3c}(x_b^i) = x_c^i$$

$$i + 1 \longrightarrow i$$

Figure 5.15
Solution by iteration using numerical substitution

We can gain some insight into what makes an iterative process converge by looking at a simple set of linear equations involving a single loop.

$$a_{11}x_a + a_{12}x_b \qquad\quad = b_1$$
$$\qquad a_{22}x_b + a_{23}x_c = b_2$$
$$a_{31}x_a \qquad\quad + a_{33}x_c = b_3$$

Let us assume an output set $1b$, $2c$, $3a$.   If we use $x_a + \epsilon_a^1$ as an initial estimate for x_1, solve equation 1 for $x_b + \epsilon_b^1$, equation 2 for $x_c + \epsilon_c^1$, and equation 3 for $x_a + \epsilon_a^1$, we can compare the size of ϵ_a^2 to ϵ_a^1 to see whether the iteration is converging or diverging. (Note: the superscript on ϵ refers to iteration, not power.)

With some simple algebra we can see that:

$$\epsilon_a^n = E\epsilon_a^{n-1}$$

where

$$E = \left(-\frac{a_{1a}}{a_{1b}}\right)\left(-\frac{a_{2b}}{a_{2c}}\right)\left(-\frac{a_{3c}}{a_{3a}}\right)$$

If the absolute value of E is less than one, the size of ϵ_a decreases with each iteration and it converges. Conversely, if the absolute value of E is greater than one, the iteration diverges. If E equals one, the determinant is zero.

Here we can make an interesting observation. Each factor of E is the negative of the input coefficient over the output coefficient for one equation. Thus, if we reverse the direction of the iteration, so the output set is $1a$, $2b$, $3c$, the roles of input and output of each equation are reversed. E is now replaced by its reciprocal. Thus, reversing the direction of iteration turns divergence into convergence, and convergence into divergence.

This observation is very simple. Unfortunately, things become more complicated when there is more than a simple loop involved, or the system is nonlinear [See Steward:62].

The process of elimination may be considered a substitution of symbolic expressions. The equations are traversed once using symbolic substitution to obtain one or more sets of one equation in one unknown; see Figure 5.16. These unknowns, in this case just x_c, can then be solved directly for their numerical values. Now, beginning with these values, e.g., x_c, we proceed through the process of numerical substitution shown in Figure 5.15 to obtain the numerical values of the remaining variables. But this time because we began with known values instead of guesses, no further iteration is required.

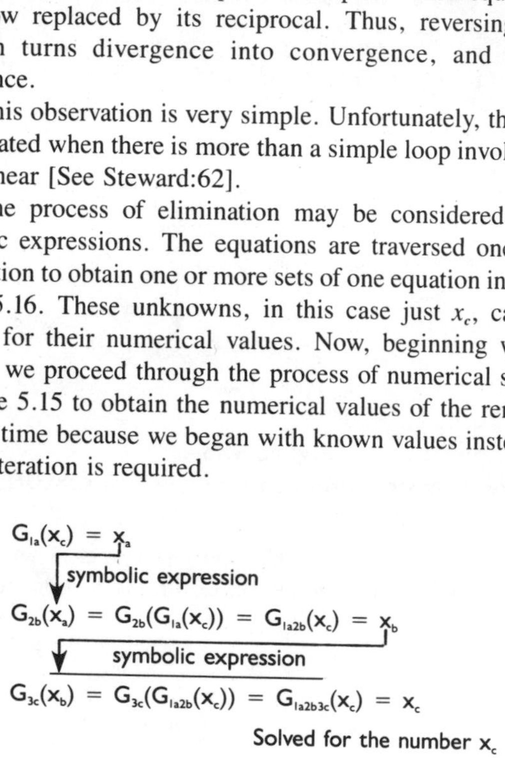

$$G_{1a}(x_c) = x_a$$
$$\downarrow \text{symbolic expression}$$
$$G_{2b}(x_a) = G_{2b}(G_{1a}(x_c)) = G_{1a2b}(x_c) = x_b$$
$$\downarrow \text{symbolic expression}$$
$$G_{3c}(x_b) = G_{3c}(G_{1a2b}(x_c)) = G_{1a2b3c}(x_c) = x_c$$
$$\text{Solved for the number } x_c$$

Figure 5.16
Solution by elimination using symbolic substitution

For both iteration and elimination, given an output assignment and an ordering of the equations, the marks above the diagonal (feedbacks) play a special role. In iteration they represent the use of estimated numerical values for variables to begin the iteration. In elimination they represent variables carried symbolically until a single equation for each such variable can be obtained and solved. The process of tearing, then, pertains to choosing which marks will play this special role and the order in which the iteration or elimination is to be performed.

5.4 Tearing Systems of Equations

We have already proved that we would get the same partition of a system of *n* equations in *n* unknowns no matter what we chose as the output assignment. Now we ask, would the arcs that can be torn and their effect on breaking circuits be the same for any choice of output assignment? As we shall see below, the answer is no. The effect of the choice of output assignment on the selection of feedback tear sets is the concern of this section.

Given a structural matrix and an output assignment, we can generate a variable precedence matrix and a set of principal circuits and shunt diagrams. We then ask how these shunt diagrams would change if the output assignment were changed. We shall show specifically how the shunt diagram changes for a relabeling of the output around the principal circuit.

We demonstrated that we are constrained in changing from one output assignment to another to move the assignment around loops. We shall capitalize upon this constraint in this analysis.

But first, let us look at the effect on arcs entering and exiting a circuit corresponding to a loop around which we relabel the output.

Figure 5.17a is a structural matrix. Let us assign the outputs indicated by the circled x's on the diagonal. The row labels for the variable precedence matrix would be as given down the left side of the matrix. The graph would be as given in 5.17b. Let p be the circuit (a,b,c,d,a). The arc (a,e) will be said to *exit* the circuit p since its first vertex is in the circuit p. The arc (f,d) will be said to *enter* the circuit p since its second vertex is in p.

Now we relabel the output around the loop in the structural matrix corresponding to the circuit p. The new outputs are indicated by boxes in 5.17c. The row labels of the variable precedence matrix now would be as given down the right side of the matrix. The graph corresponding to this precedence matrix is shown in 5.17d. Shifting outputs to the diagonal gives the precedence matrix in 5.17e.

Now we can make the following observation:

OBSERVATION 5.2.   Given a variable precedence matrix and its graph, relabeling the output assignment around a loop corresponding to a circuit p produces a new variable precedence matrix and graph with the following changes:

1.   The output marks along the loop in the structural matrix corresponding to p have become nonoutput marks, and the nonoutput marks along the loop have become output marks.

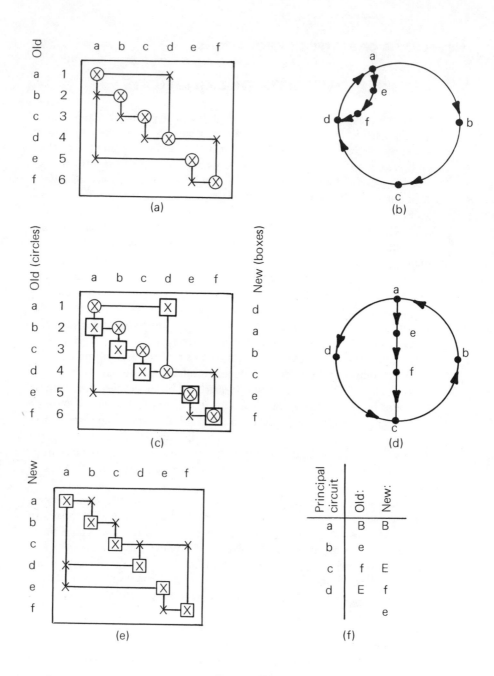

Figure 5.17
(a) Structural matrix with old output assignment on diagonal—thus also the variable precedence matrix. (b) Graph for old output assignment. (c) Structural matrix with old output assignment (circles) and new output assignment (boxes). (d) Graph for new output assignment. (e) Variable precedence matrix for new output assignment. (f) Shunt diagram: read down for old output assignment, read up for new output assignment.

2. The labeling of the vertices in the circuit p of the graph are not changed, but the direction of p is reversed.
3. The labeling of a vertex of an arc is not changed if that vertex is not on p.
4. If a vertex is the first vertex of an arc which exits from p, then its labeling does not change.
5. If the vertex is the second vertex of an arc entering p, then its identification is changed to the identification of the preceding vertex in the original orientation of p.

For example, we can see that the relabeling in Figure 5.17c does not change the identification of the vertices in the exiting arc (a,e), but does change the entering arc from (f,d) to (f,c).

With this observation we can infer how shunt diagrams will be changed by relabeling of the output assignment around the loop corresponding to the principal circuit. Then we can show how we might sometimes wish to relabel to obtain better opportunities for tearing. The consequences of relabeling the outputs around circuits other than the principal circuit can similarly be derived by applying Observation 5.2.

Figure 5.17f shows the shunt diagrams before and after relabeling the output around the principal circuit (a,b,c,d,a). To compare the shunts before and after the relabeling, we have shown the new shunts without rewriting the principal circuit. Before the relabeling, the principal circuit is read from top to bottom cyclically. After the relabeling, the principal circuit is read from bottom to top cyclically.

This leads to the following observation:

OBSERVATION 5.3. Given a variable precedence matrix and a shunt diagram with the principal circuit p, let the output be relabeled around the loop corresponding to the principal circuit p. Before the relabeling, we read the principal circuit from top to bottom cyclically. After the relabeling, we will read the principal circuit from bottom to top cyclically. Then for each shunt after the relabeling:

1. The B stays in the same position.
2. The E moves one position backward in the original orientation of the principal circuit. This shortens by one the length of the shunted path in the principal circuit.
3. The shunt proceeds from B to E, bottom to top cyclically.
4. The roles of the fences and gaps become interchanged, i.e., the gaps become fences and the fences become gaps.

Thus, instead of looking only for arcs in the principal circuit which have few or no shunts parallel to them, we now look also for arcs in the principal circuit which have many shunts parallel to them. Although such a choice of arcs may be a poor choice before the relabeling, after the relabeling, the fences become gaps and it may become a good choice. For example, looking at parts (b), (d), and (f) of Figure 5.17, we can see that before the relabeling tearing either arc (a,b) or arc (b,c) or arc (c,d) would leave the parallel shunt (a,e,f,d). But after the relabeling tearing either arc (b,a) or (c,b) would leave no parallel shunt.

Note that if the equation precedence matrix were used to obtain the shunt diagram, instead of the variable precedence matrix used here, the effect of reversing the direction of the principal circuit would be to first move the B (rather than the E) to shorten the length of the shunted path, then to reverse the roles of fences and gaps.

5.5 Kron's Concept of Tearing

Although the concept of tearing considered here differs somewhat from the one promoted by Gabriel Kron, since our early work was motivated by his ideas, it is appropriate to discuss his method and its relation to what is discussed in this book.

Kron deserves to be called the father of tearing. He was certainly one of the earliest if not the earliest active proponent of solving complex problems by breaking the problem into pieces, solving the pieces, then assembling these solutions into the solution of the whole. He called it diakoptics [Kron:63]. Unfortunately, the understanding and appreciation of his method was sparse and scattered. His personal enthusiasm for his ideas and his estrangement from engineers who thought him too much of a mathematician, and from mathematicians who thought him too much an engineer, contributed to his becoming as colorful and controversial as his ideas. Several authors have tried to translate his methods so they would be accessible to a wider audience. However, his ideas were difficult to translate because they depended so much on his own talents in choosing where to tear. The art is to find where to tear a system so that the parts are easy to solve without making it too difficult to assemble the solutions. Kron's successes in applying his methods astounded more than they revealed how to use his methods.

Kron used electric circuits as analogies in which to see where to tear. The electric circuit analogy would be described by the same equations as the system he was studying. Tearing in the circuit analogy involved cutting the circuits into subcircuits by cutting branches. The subcircuits were solved subject to unknown currents at the torn branches. These solutions were used to assemble the solution for the whole problem by solving for the currents at the torn branches. He could see the appropriate places to tear most easily by looking at the topology of the circuit.

In this book we have used shunt diagrams to see where to tear, and have shown how to use a computer program to develop these shunt diagrams.

Kron's method involves two aspects: (1) where to tear to separate the problem into smaller problems, and (2) how the solution of each of the smaller problems can be put together to obtain the solution of the original problem. We will discuss first the basis of Kron's method for solving the pieces and putting these solutions together, and then show how our methods can be used to see where to tear to get the pieces.

Bueckner (1956) and Householder (1960) have shown that when Kron's method for assembling solutions is put into matrix form it corresponds to methods which had also been developed by Woodbury (1950) and by Sherman and Morrison (1949, 1950). (See also Householder[53].)

Tearing as done by Kron and revealed by others concerned linear systems. (The methods developed in this book do not depend upon the system being linear.)

The matrix formulation of Kron's method is as follows: Assume we want to solve the following system of linear equations:

$A = B + KL$ where

B is an $n \times n$ matrix which is easy to solve

K is an $n \times r$ matrix

L is an $r \times n$ matrix

Frazer et al. (1953) shows that if $A - B$ has order n and rank r, it can always be represented by an $n \times r$ matrix (K) times an $r \times n$ matrix (L).

Now we shall show that we can solve for x by solving a system of equations using B and solving another system of r equations. If the system of equations B is indeed easy to solve and r is small as compared to n, then this method may be easier to use than solving the full system by more direct means. We will thereafter consider how to use the methods of tearing developed here to help break A into B, K, and L.

Let us assume:

$$Ax = b$$
$$(B + KL)x = b$$
$$Bx = b - KLx$$
$$x = B^{-1}b - B^{-1}KLx$$

Let $y = Lx$ and note that y is an $r \times 1$ vector. Then,

$$y = Lx = LB^{-1}b - LB^{-1}Ky \quad \text{or}$$
$$y = (I + LB^{-1}K)^{-1}LB^{-1}b \quad \text{and}$$
$$x = B^{-1}b - B^{-1}K(I + LB^{-1}K)^{-1}LB^{-1}b$$

Note that $(I + LB^{-1}K)$ is a system of r equations. Thus we can indeed solve for x by solving the system B and a system of r equations, provided we can show that $(I + LB^{-1}K)$ has a nontrivial solution. We demonstrate this by showing that given any $r \times 1$ vector z

$$(I + LB^{-1}K)z = 0$$

implies that z must be zero. Multiplying on the left by $BB^{-1}K$ we have

$$(B + KL)B^{-1}Kz = 0 \quad \text{or}$$
$$AB^{-1}Kz = 0$$

Since A and B are nonsingular, it must be that $Kz = 0$. Substituting this into the first equation above involving z shows that $z = 0$, as we wished to prove.

Let us look at the final equation for x and show the number of rows and columns in each of the matrices and vectors as follows:

$$x_{n\times1} = B^{-1}_{n\times n} b_{n\times1} - B^{-1}_{n\times n} K_{n\times r}(I_{r\times r} + L_{r\times n}B^{-1}_{n\times n}K_{n\times r})^{-1}_{r\times r}L_{r\times n}B^{-1}_{n\times n}b_{n\times1}$$

We note that the solution for x is obtained by solving the system with the matrix B, then adding a correction term. The correction term requires the solution of a system of r equations, where by assumption r is much smaller than n.

Now, let us consider how the methods of this book help determine how to break A into B, K, and L. Tearing a system of equations orders the equations and variables so that we have a block triangular system plus certain tear elements above the block diagonal. The block triangular system becomes the B matrix, while the tear elements become the KL. If the B matrix were block diagonal, the blocks would represent subsystems which can be solved one at a time. If B is block triangular, then the subsystems represented by the blocks can be solved one at a time, but the solution of one block may depend upon the solution of other blocks that have already been solved. The KL represents the coupling of these solutions to obtain the solution of the original problem.

It can be seen easily that r, the number of columns of K and rows of L, can be made to be the smaller of the number of columns or the number of rows that contain all the tears. Section 4.2 discussed the use of shunt diagrams to obtain tears in a small number of rows or columns.

A small problem will illustrate the method. Obviously, the larger the problem the more useful the method, but also the more difficult it is to present as an example.

Figure 5.18 shows the system of equations. Figures 5.19 through 5.22 show the application of the techniques of this chapter and section 4.2

$$
\begin{aligned}
y_b + 2y_c & = -1 \\
y_a \qquad\qquad - y_d + 2y_e & = -3 \\
y_a \qquad\qquad\qquad\qquad + y_f & = 0.5 \\
-y_b \qquad\qquad + y_e + y_f & = 1 \\
2y_a + 3y_b \qquad\qquad + 6y_e & = 1 \\
2y_c + 2y_d \qquad\quad + y_f & = 2
\end{aligned}
$$

Figure 5.18
Equations

	a	b	c	d	e	f
1		⊗	x			
2	x			⊗	x	
3	x					⊗
4		x			⊗	x
5	⊗	x			x	
6			⊗	x		x

Figure 5.19
Structural matrix with output
assignment

		a	b	c	d	e	f
a	5	⊗	x			x	
b	1		⊗	x			
c	6			⊗	x		x
d	2	x			⊗	x	
e	4		x			⊗	x
f	3	x					⊗

Figure 5.20
Equations reordered to bring outputs
to diagonal

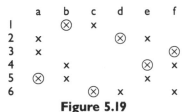

5	a	B	B		E
2	d	f	f	E	
6	c	E			
1	b				B
4	e		E	B	

Figure 5.21
Shunt diagram with tears in a small
number of rows. Tear 4b and f, 5b

		e	a	d	f	c	b
4		⊗			x		x
5		x	⊗				x
2		x	x	⊗			
3			x		⊗		
6				x	x	⊗	
1						x	⊗

Figure 5.22
Equations and variables reordered
to show tears above the diagonal

to order the equations to reveal the tearing. The original equations with rows and columns reordered as per Figure 5.20 are as follows:

$$
A = \begin{vmatrix} 1 & 0 & 0 & 1 & 0 & -1 \\ 6 & 2 & 0 & 0 & 0 & 3 \\ 2 & 1 & -1 & 0 & 0 & 0 \\ 0 & 1 & 0 & 1 & 0 & 0 \\ 0 & 0 & 2 & 1 & 2 & 0 \\ 0 & 0 & 0 & 0 & 2 & 1 \end{vmatrix} \quad y \quad b = \begin{vmatrix} 1 \\ 1 \\ -3 \\ 0.5 \\ 2 \\ -1 \end{vmatrix}
$$

$$A = B + KL$$

where

$$
B = \begin{vmatrix} 1 & 0 & 0 & 0 & 0 & 0 \\ 6 & 2 & 0 & 0 & 0 & 0 \\ 2 & 1 & -1 & 0 & 0 & 0 \\ 0 & 1 & 0 & 1 & 0 & 0 \\ 0 & 0 & 2 & 1 & 2 & 0 \\ 0 & 0 & 0 & 0 & 2 & 1 \end{vmatrix} , \quad KL = \begin{vmatrix} 0 & 0 & 0 & 1 & 0 & -1 \\ 0 & 0 & 0 & 0 & 0 & 3 \\ 0 & 0 & 0 & 0 & 0 & 0 \\ 0 & 0 & 0 & 0 & 0 & 0 \\ 0 & 0 & 0 & 0 & 0 & 0 \\ 0 & 0 & 0 & 0 & 0 & 0 \end{vmatrix}
$$

Then it can be seen that

$$
K = \begin{vmatrix} 1 & 0 \\ 0 & 1 \\ 0 & 0 \\ 0 & 0 \\ 0 & 0 \\ 0 & 0 \end{vmatrix} , \quad L = \begin{vmatrix} 0 & 0 & 0 & 1 & 0 & -1 \\ 0 & 0 & 0 & 0 & 0 & 3 \end{vmatrix}
$$

Now we can compute the components of the equation for x as follows:

$$
B^{-1}b = \begin{vmatrix} 1 \\ -2.5 \\ 2.5 \\ 3 \\ -3 \\ 5 \end{vmatrix} , \quad B^{-1}K = \begin{vmatrix} 1 & 0 \\ -3 & .5 \\ -1 & .5 \\ 3 & -.5 \\ -0.5 & -.25 \\ 1 & .5 \end{vmatrix}
$$

$$I + LB^{-1}K = \begin{vmatrix} 3 & -1 \\ 3 & 2.5 \end{vmatrix}, \qquad (I + LB^{-1}K)^{-1} = \frac{1}{10.5} \begin{vmatrix} 2.5 & 1 \\ -3 & 3 \end{vmatrix}$$

$$LB^{-1}b = \begin{vmatrix} -2 \\ 15 \end{vmatrix}$$

$$x = \quad B^{-1}b \; - \; B^{-1}K(I + LB^{-1}K)^{-1}LB^{-1}b$$

$$x = \begin{vmatrix} 1 \\ -2.5 \\ 2.5 \\ 3 \\ -3 \\ 5 \end{vmatrix} - \frac{1}{10.5} \begin{vmatrix} 1 \\ -3 & .5 \\ -1 & .5 \\ 3 & -.5 \\ -0.5 & -.25 \\ 1 & .5 \end{vmatrix} \begin{vmatrix} 2.5 & 1 \\ -3 & 3 \end{vmatrix} \begin{vmatrix} -2 \\ 15 \end{vmatrix}$$

$$x = \begin{vmatrix} 1 \\ -2.5 \\ 2.5 \\ 3 \\ -3 \\ 5 \end{vmatrix} - \frac{1}{10.5} \begin{vmatrix} 10 \\ -4.5 \\ 15.5 \\ 4.5 \\ -17.75 \\ 35.5 \end{vmatrix} = \begin{vmatrix} .047619 \\ -2.071429 \\ 1.023810 \\ 2.571429 \\ -1.309524 \\ 1.619049 \end{vmatrix}$$

Now y is the vector x with a reordering of the elements. Using the columns in Figure 5.22 to restore this ordering we have:

$$y = \begin{vmatrix} -2.071429 \\ 1.619049 \\ -1.309524 \\ 1.023810 \\ 0.047619 \\ 2.571429 \end{vmatrix}$$

6 Critical Path Scheduling

6.1 What Is a Critical Path Schedule?

The fast, competitive pace of modern-day technology requires bold commitments to projects before there is complete knowledge of all the circumstances which may arise. Customer requirements, government regulations, the state of knowledge due to development or testing, the competitive situation, strikes, delays, accidents, etc., can continually introduce changes into the project as it proceeds. We can usually expect that there will be unforeseen events. Unfortunately, we do not know beforehand what these unforeseen events will be, but when they do occur we must have the ability to determine rapidly their consequences so that appropriate action can be taken.

In the late 1950s, a scheduling technique designed to cope with this challenge was developed. This basic technique is called critical path scheduling, CPS. At about the same time, two groups were responsible for developing variations on this basic concept. Remington-Rand and DuPont developed jointly a means of planning the best cost schedule when given the costs of expediting each activity. This variation is frequently called the Critical Path Method, CPM. A group at Booze, Allen and Hamilton developed for the U.S. Navy Special Projects Office a variation called PERT, Project Evaluation and Review Technique. PERT added to the basic CPS concept the notion of the probability of meeting the schedule as a function of the uncertainty in the time required to complete each of the activities in the project.

Although the use of probability in PERT seemed to be a good idea at the time, the method used to compute the probabilities was based upon an invalid assumption. Thus, if management were to believe the probabilities computed by PERT, they may be led into making some very costly management errors. We shall discuss the use of probabilities later in this chapter.

CPM has been very successfully used in the construction business. PERT was credited, in large measure, for the Polaris Missile Project's being completed two years ahead of schedule. The Department of Defense, NASA, and various civilian agencies have now brought their contracts under some form of critical path control. The dramatic accomplishments due to these tools were such as to demand wide management attention.

Network or critical path scheduling techniques such as PERT or CPM are now well established and proven tools. This chapter deals with an introduction of the basic concepts of critical scheduling, and a discussion of and critique of how these tools can be improved upon further. We shall introduce some new critical path techniques and lay the foundations for the use of critical path in conjunction with the design structure system to be discussed in the next chapter.

Critical path scheduling is a good place to start building a management information system. It can start small and grow. Critical path scheduling can be introduced into one part of the organization without upsetting the existing institutions. It can begin with a manager who wants to try it; other managers will become interested when they see how its use can protect their interests. As its use grows, it lays the paths of data flow which form the base upon which more elaborate management information systems can be built. As it grows it can be assimilated into the existing system. The experience gained in the use of this tool can be used to guide its improvement and extensions which will evolve as the needs of the concern and management's experience with the tool develop. Such extensions are discussed in this chapter.

The initial use of the scheduling tool is to evaluate trial plans. Several plans may be evaluated before the project begins. Once the project is underway, as actual beginning and finish dates, various troubles, and changes in estimates are reported, the scheduling tool will then be used to follow the implications of any change from the original plans and show where replanning might be necessary. The principal advantage of such formal scheduling tools is not to develop schedules where the plans can remain fixed, but where plans must be changed because of unforeseen circumstances, which are always to be expected, though we may not know in what form.

Through the scheduling tool, the effect of a delay or trouble in one phase of the project will be communicated to all those managers whose operations will be affected. For example, let us assume that in the original plan made in January, the delivery of part A on September 1 is critical to the completion of the project on time. The purchasing agent must pay the vendor a premium to cover his overtime to deliver part A by this date. Now let us assume that in June a delay is incurred in the process to manufacture part B which must mate with part A. This means that part A cannot be used on September 1 as originally expected. Manufacturing reports the delay to the scheduling system. In updating the schedule on the computer, the effect on all other activities is computed. The resulting report that goes to the purchasing agent shows that part A is now not needed until November 15. Confirming this by phone, he can modify the arrangments with the vendor to save money. If the purchasing agent were not made aware of this delay in manufacturing, he would find himself continuing a premium to expedite delivery of a part which then accrues an inventory charge while waiting to be used.

CPS is a system in which managers responsible for each part of the system make their report into the central system, and the implications on the schedule are computed and reported to those managers responsible for the activities affected.

To keep ahead of the situation, it is necessary that the maintenance of the input data not be overburdening. The scheduling tool should be designed to minimize red tape and manual data handling required to supply the tool with up-to-date information, i.e., actual times of beginning and finish of activities, problems, and changes of estimated time and plans.

We use a method of describing the project network to the scheduling program which is simpler and requires less data handling than did the original PERT and CPM. The scheme takes the burden of maintaining diagrams, event numbers, and dummy activities off the user; this can greatly increase the responsiveness of the tool (the comparison with PERT is discussed in section 6.7.)

6.2 How a Project Is Described

We shall begin with a discussion of a simple scheduling tool without consideration of personnel and resource allocation. Only one time estimate will be used here. (A critique of the three-time estimate system used in PERT is made in section 6.10.)

To describe a project, the project is broken down into activities. For each activity we consider (1) which other activities must be finished before this activity can be started (i.e., the predecessor activities), and (2) how long it will take to perform this activity (i.e., duration). For example, a project may be described by a table as in Table 6.1. The activities and predecessors are chosen such that an activity can begin only when all of its predecessors are finished. In a later section we shall relax this restriction. For now, though, this table gives the basic information required to develop a schedule.

The predecessors may be represented by a network diagram as in Figure 6.1. Each circle represents an activity to be performed; the lines indicate the precedences. For example, the line from circle 2 to circle 6 indicates that activity 6 cannot be started until activity 2 is complete. The durations are written below the circles.

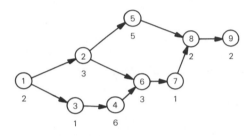

Figure 6.1
Network Precedence Diagram

Given a beginning time for the project, then either the table or diagram can be used to compute a schedule. We state the rules first, then apply them to an example which will make the rules clearer.

Table 6.1: PROJECT DESCRIPTION

Activity Number	Activity Description	Duration	Predecessors
1	Design	2	None
2	Drafting	3	1
3	Prepare purchase orders	1	1
4	Awaiting delivery	6	3
5	Prepare manuals	5	2
6	Assemble	3	2, 4
7	Check out	1	6
8	Transport	2	5, 7
9	Field test	2	8

6.3 Rules for Computing Earliest Times

The earliest beginning and finish times are the earliest times that an activity can be started or finished such that all the required predecessor activities are finished as early as possible:

1. If an activity is not preceded by any other activity, its earliest beginning time is the beginning time of the project.
2. The earliest finish time of an activity is the earliest beginning time plus the duration.
3. The earliest beginning time of an activity which has predecessors is the largest of the earliest finish times of its predecessors.
4. The earliest finish time of the project is the largest of the earliest finish times of those activities which have no successor.

We begin with those activities with no predecessors. We then compute the earliest beginning and finish times in such an order that when an activity is computed, all its predecessors have already been computed. Any order which satisfies this condition is called in critical path parlance a *topological order*. Procedure 3.1 showed how to obtain such a topological ordering. In this way we compute all of the earliest times starting at the beginning of the project and moving toward the end. Clearly, such a topological order cannot exist if the precedences imply a circuit, e.g., *a* precedes *b*, and *b* precedes *a*. In the fabrication of systems, circuits should not occur. The next chapter will discuss circumstances in design and engineering where circuits do arise, and will develop tools which can be used in conjunction with critical path techniques to deal with them.

6.4 Rules for Computing Latest Times

The latest beginning and finish times are the latest times an activity can be started or finished without delay of the project:

1. If an activity is not succeeded by any other activity, its latest finish time is taken as the required completion time for the project.
2. The latest beginning time is its finish time minus its duration.
3. The latest finish time of an activity which has successors is the smallest of the latest beginning times of the activities which succeed it.

6.5 Rules for Computing Slacks

The slack is the length of time an activity can be delayed from its earliest time without delaying the project.

The slack for an activity is its latest beginning time minus its earliest beginning time, or, equivalently, its latest finish time minus its earliest finish time.

We sometimes prefer to call this concept of slack, *shared* slack, to serve as a reminder that when an activity is delayed it may decrease the slack of other activities also.

6.6 An Annotated Example

The calculations can be made in the form of a table, or they can be done directly on the diagram.

In Figure 6.2 the earliest beginning and finish times are computed, working from left to right in the order of the numbers within the circles (these numbers are in topological order). We are told the project begins at time 5, so we write 5 for the earliest beginning time for activity 1. The times are written around the circles as indicated by the key. Two units of time are required to perform activity 1, so its earliest finish time is $5 + 2 = 7$. We proceed similarly for activities 2, 3, 4, and 5. Activity 6 is

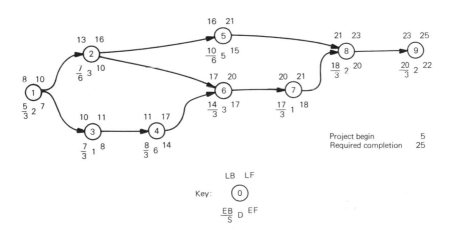

Figure 6.2
Network Precedence Diagram with Computed Times

preceded by two activities and cannot begin until both are finished. Thus, the earliest beginning time of activity 6 is the greater of 10 and 14. Proceeding in this way, we finally compute the earliest finish time for the last activity of the project as 22. We then use the project required finish time of 25 as the latest finish for the last activity (or activities) and work from right to left computing latest times. If activity 9 has a latest finish time of 25, then the latest it could begin is $25 - 2 = 23$. If activity 9 must begin by time 23, then activity 8 must be finished by time 23. Activity 2 has two successors. If neither successor is to be delayed, then activity 2 must be finished in time for the earliest successor, i.e., at time 16. After the earliest and latest times have been computed, the slack can be computed as the difference between the earliest and latest beginning times. The same calculations can be done in Table 6.2.

Note that there is a sequence of activities extending from the beginning to the end of the project with a minimum slack, in this example, 3. The longest sequence of activities has this minimum slack. This path, which controls the completion time of the project, is called the critical path. Clearly, it is these minimum slack, or critical path, activities to which we must pay first attention if the schedule is to be met or expedited.

TABLE 6.2: SCHEDULE CALCULATION

Activity Number	Description	Duration	Predecessors	EB	EF	LB	LF	Slack
1	Design	2	None	5	7	8	10	3
2	Drafting	3	1	7	10	13	16	6
3	Prepare purchase orders	1	1	7	8	10	11	3
4	Awaiting delivery	6	3	8	14	11	17	3
5	Prepare manuals	5	2	10	15	16	21	6
6	Assemble	3	2, 4	14	17	17	20	3
7	Check out	1	6	17	18	20	21	3
8	Transport	2	5, 7	18	20	21	23	3
9	Field test	2	8	20	22	23	25	3

6.7 Comparison of Precedence Networks with Activities on the Lines Networks

Those readers familiar with the original PERT or CPM should note that in the method we used, it was not necessary to draw a diagram to assign event numbers and dummy activities before we or the computer ran the calculation of the schedule. Sometimes one wants to draw a PERT chart where the length of the activity lines represents the duration of the activity. If we wish to draw the conventional PERT chart with the activities on the lines, the next section shows how we can have the computer assign the event numbers and dummy activities so we can draw or have the computer draw the diagram without any further manipulation. Let us consider the following example to see some of the burden and possible errors which we avoid by having the computer do this; see Table 6.3.

Note that the precedence table can be filled out directly by considering what needs to be finished before an activity can start. This table is sufficient to go directly as input to the computer. Management is then completely familiar with the input to the computer program that will generate the information upon which decisions will be based.

TABLE 6.3:   Duration and Precedence Table

Activity Number	Estimated Duration (weeks)	Predecessors
1	3	None
2	3	1
3	5	None
4	1	1, 3

PERT programs, which use activity on the line networks, require a different form of input to convery the precedences. Let us see how we must prepare the aforementioned precedence data as input to a PERT program.

The input to a PERT program implies the precedences by the use of event numbers assigned to the beginning and end of each activity. In the conventional scheme, a PERT diagram must be drawn with the activities represented by lines to establish these event numbers. For the above examples, we would be inclined to draw a PERT diagram as in Figure 6.3.

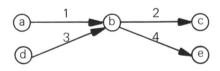

Figure 6.3
PERT Diagram

We see that this diagram implies all of the required precedences, and thus we might proceed to prepare PERT precedence input as in Table 6.4.

TABLE 6.4:   PERT INPUT

Activity	Beginning Event	Final Event
1	a	b
2	b	c
3	d	b
4	b	e

However, we note on more careful observation that this diagram and the PERT input imply a predecessor which management did not intend, namely, that 3 precedes 2. This superfluous predecessor has the effect of extending the computed project completion from 6 weeks to 8 weeks.

The difficulty arose from activity 3, which we intend to precede 4 but not 2. But we made the beginning events of activities 2 and 4 common, so that everything which precedes the one must then precede the other.

To avoid this difficulty in conventional PERT, we must represent the beginnings of activities 2 and 4 as separate events. We introduce a dummy activity with no duration to imply their common predecessor, as seen in Figure 6.4.

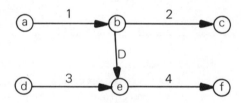

Figure 6.4
Corrected PERT Diagram

Tracing through this diagram, we can now verify that the precedences are precisely those we implied in Table 6.3—no less, no more. Then we can make up our PERT precedence input as shown in Table 6.5.

TABLE 6.5:   PERT INPUT—CORRECTED

Activity	Beginning Event	Final Event
1	a	b
2	b	c
3	d	e
4	e	f
D	b	e

To emphasize the problem still further, let us assume that after the project is begun it is realized that another activity must be performed before activity 2 can be started. Thus, we wish to add an activity 5 preceding activity 2. If we were to use the PERT form of input, we would find that to add this activity requires the introduction of an additional dummy activity, so the new PERT chart must be redrawn as in Figure 6.5. The PERT input must then be modified as in Table 6.6, where the changes are underlined.

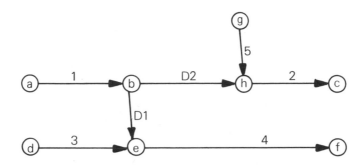

Figure 6.5
PERT Diagram with Added Activity

TABLE 6.6: PERT INPUT WITH ADDED ACTIVITY

Activity	Beginning Event	Final Event
1	a	b
2	h	c
3	d	e
4	e	f
5	g	h
D1	b	e
D2	b	h

Putting the activities on the vertices rather than the arcs also allows overlaps between a task and the predecessor. See section 7.4 and Figure 7.3.

The maintenance of large PERT charts involving hundreds or thousands of activities is no trivial task. Often the schedule suffers because of the difficulty and delay involved in the intermediate step of maintaining the PERT diagrams required to maintain the computer input. With the method we have used we would have the precedence table as in Table 6.7.*

*The author tried to publish a paper in 1962 proposing that the precedence method gets rid of the dummies. From the context in which this remark was made, it appeared he was proposing to get rid of not only the dummy activities but the people who had to draw them. The paper was rejected and the proposal considered impertinent. Therefore, the scheduling program based on this method has been called IMPERT.

**TABLE 6.7: DURATION AND PRECEDENCE
TABLE WITH ADDED ACTIVITY**

Activity Number	Estimated Duration (weeks)	Predecessors
1	3	None
2	3	1,5
3	5	None
4	1	1, 3
5	4	None

6.8 Rules for Converting Precedence Networks to Activities on the Lines Networks

This section deals with rules that can be programmed into a computer scheduling system. It may be skipped by the reader who intends only to be a user of such a program.

The rules can be executed by a computer program so that a PERT-type schedule with the activities on the lines can be plotted by a computer-driven plotter. The length of the lines for the activities can be plotted to be proportional to the durations of the activities. It is particularly desirable to have this done on a computer if there are constant changes and revisions which would cause the chart to be redrawn continually.

We begin by considering each activity as having its own beginning and end event, with all the precedences indicated by dummy activities. Then we determine which events can be coalesced and which dummies are thereby removed. For example, given the precedence table in Table 6.8, consider the diagram in Figure 6.6.

The rules, and their application to the example, are as follows:

1. Two or more end events may be coalesced if they are succeeded by precisely the same set of non-dummy activities (Figure 6.7).
2. Two or more beginning events may be coalesced if they are preceded by precisely the same set of non-dummy activities (Figure 6.8).
3. The events at either end of a dummy may be coalesced, if either (a) the left event of the dummy has no other activity following it, or (b) the right event of the dummy has no other activity preceding it (Figure 6.9).

TABLE 6.8: PRECEDENCE TABLE

Activity Number	Predecessors
1	None
2	1
3	1
4	None
5	None
6	1, 4, 5

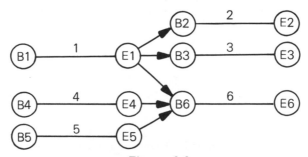

Figure 6.6
Activities on Lines and Dummies for Predecessors

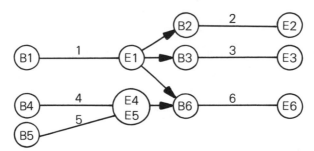

Figure 6.7
Rule 1

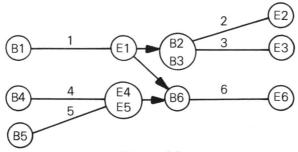

Figure 6.8
Rule 2

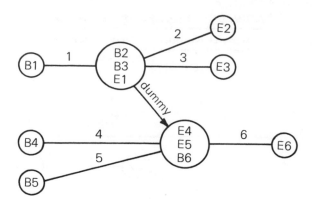

Figure 6.9
Rule 3

4. By convention, dummies are usually introduced as necessary so that no two activities have both the same B and E events (Figure 6.10).

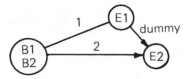

Figure 6.10
Rule 4

6.9 The Use of Events

An event is a point in time (i.e., one date), while an activity normally represents an interval of time. Events may be incorporated into a precedence network as activities with zero duration. Such activities representing events can be added to the network for the following purposes:

Events as milestones: To obtain a schedule, the project plan must be laid out with all of its activities. However, higher levels of management are usually not interested in the detailed activities, but only in the completion of certain major milestones or events. Higher level reporting may be done by events rather than activities, or possibly both.

Events for network integration: Some projects involve several thousand activities. Handling a single network with this many activities can become clumsy and the interpretation quite difficult. It is often desirable to break the total project into many subnets which can then be integrated to obtain the same schedule as would be obtained if all the activities were in one network.

Frequently, these subnets can be associated with distinct parts of the project, or distinct contractors or subcontractors. Events can be used for interfaces between subnets. In each subnet, the critical path is determined from each initial interface event to each final interface event that follows. In the higher level network, the subnet is replaced by the interface events, with activities between them of duration equal to the length of the critical path between these events as they appear in the subnet. Once these event times are established in the higher level network, these times can be placed as constraints on the same events in the subnet to schedule the detailed activity beginning and finish times in the subnet.

It sometimes occurs that there are several activities which all have the same set of predecessors. To save having to reiterate these predecessors for each such activity, it may be useful to introduce an event having the common set of predecessors. When such a condition occurs, the event usually represents the completion of one phase and the beginning of another phase of the project.

6.10 Schedule Uncertainty and the Misuse of Probability

PERT computes the uncertainty in the scheduled finish time for the project, given the uncertainties in the durations of the activities. This information could be very useful. However, the method used by PERT for many years and in many important projects to compute probability distributions is not correct. The error is such that it could lead the user into serious decision errors. Fortunately, most mature PERT users, for one reason or another, have learned to avoid the use of this feature.

PERT uses three estimates for duration: t_e = an optimistic time such that there is only a one percent chance that the activity can be finished sooner; the pessimistic time, t_l, is such that there is only a one percent chance that the activity will be finished later; and t_m = a most likely time. These three times imply a probability distribution for the duration. PERT then takes as the expected duration $(t_e + 4t_m + t_l)/6$, and as the variance

$[(t_l - t_e)/6]^2$. PERT computes the median time (improperly confused with "expected" time in the PERT literature) and the variance for the project completion. The median time is such that it is equally likely the project will finish either earlier or later.

When the durations of activities in sequence are added, the median times and their variances are added. This is valid. However, when the latest time of one or more activities is computed, the variance, for simplicity, is assumed to be the single variance of the one latest time. The variances of the other paths are ignored. This is correct so long as the one latest time is significantly later than any of the other times. It is not correct if the times are about equal. This occurs frequently when the schedule has been expedited and leads to serious error which can in turn lead to bad scheduling decisions.

As a schedule is expedited, activities on the critical path are expedited first, until more and more paths become critical or nearly critical. This increases the vulnerability of the computed schedule times because it increases the number of activities whose delay will cause a delay in the project. This effect can be serious, and management should be aware of it.

The simplifying assumption made by PERT causes this effect not to be properly considered. The management using PERT could probably make their own compensations for this effect if they were not led to expect that PERT already does this for them.

It costs to expedite activities. It is important that a scheduling tool estimate properly the time saved by this costly expediting. In fact, PERT tends to overestimate the expected value of the time saved.

The distribution of possible project finish times, when properly calculated, may become very skewed. For example, the curve in Figure 6.11 conveys the idea that there is about a 50–50 chance that the project

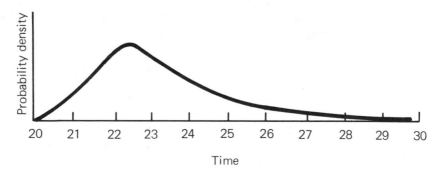

Figure 6.11
Skewed Distribution

would be finished before time 23 because about half of the area of the curve is to the left of 23. However, there is some chance that the project will be delayed as long as time 30.

Such skew can be quite important, particularly when determining the possible financial risk due to the cost of project delays. PERT implies that the curve is symmetric and does not show any skew in its computed finish time distribution.

But PERT not only doesn't show the skew, it underestimates the expected and median completion times.

The equations for the probability calculations used in PERT have been repeated over and over in the literature without pointing out the fallacy in the assumptions used.

A common objection to PERT is the need to estimate optimistic and pessimistic times. People just do not have a feeling for such quantities. However, it is not unreasonable to ask that they specify an activity as having one of a set of characteristic standard distributions and a specified percent variance. Various activities, such as purchase from stock, shop fabrication, testing, development, etc., have their own characteristic distributions. The use of such distributions is simpler than trying to estimate a unique distribution for every new activity. Furthermore, experience over a number of similar activities can be incorporated into the specification of these standard distributions.

An improved method of probability calculation which properly computes the expected time and shows the skew is reported elsewhere [Steward:67b]. This method computes the probability distributions using Laplace transforms and approximations. Others have handled this problem by doing Monte Carlo calculations.

6.11 Planning and Control

The schedule may be used for estimating the completion date of the planned project. If the completion date does not satisfy the required finish date (i.e., the minimum slack is negative), another plan may be drawn up and its completion date estimated. Perhaps by repeating some work it will be possible to perform some activities in parallel which must otherwise be done in series, or perhaps critical activities can be speeded up by assigning more people or machines.

One must be careful that every change made in the project plan is a reflection of an actual change in the way the work is done and not just the desire to meet commitments. One should be careful about decreasing the estimated duration of an activity without showing how that activity can be done in less time, e.g., assigning more people. The computed completion time can be no more meaningful than the estimates used.

A project cannot be expected to follow precisely the plan drafted at the beginning of the project. The principal advantage of these scheduling tools is the ability to see quickly where replanning is required as conditions deviate from the established plan. To partake of this advantage it is necessary to make the collection of information and updating of the schedule sufficiently convenient that frequent maintenance is quick and practical. This collection of information and input maintenance is facilitated by the use of the return form shown in the next section and by the simplified way of describing the project network.

6.12 An Example of Computer Output with Return Form

Let us consider what the output from a computer program to do these calculations might look like. (Although a computer is not necessary for critical path scheduling, it is a great convenience.)

Each manager receives a package (Figure 6.12a) consisting of:

1. The manager's mailing address
2. A project summary and detailed schedule on just those activities requested (Figure 6.12b)
3. A return form for those activities on which the manager must report (Figure 6.12c)
4. The mailing address for the return form

The package, as it comes from the computer, is stapled and put directly into the plant mail. At the end of the reporting period, the manager tears off the return form, fills in the pertinent spaces, staples it, and puts it in the mail. Or this information may be presented on a computer terminal.

Let us look at the project schedule in Figure 6.12b. In the project summary, BEGINNING (1/08/79), NOW (3/19/79), and DUE COMMITMENT (8/03/79) are as given in the input. The COMPUTED FINISH is the finish time computed by the schedule on the basis of the project beginning time, actual beginning and finish times of activities, and estimated durations. The computed finish date is subtracted from the due commitment date for comparison, i.e., DUE FINISH -9.4 indicates the project is computed by the schedule to finish 9.4 weeks later than the

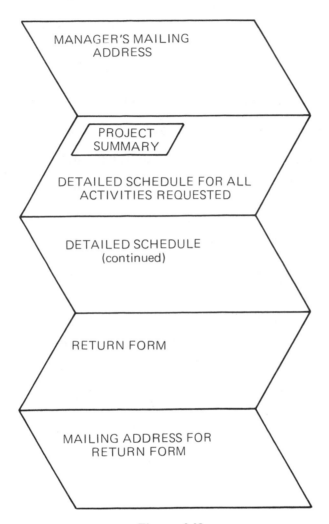

Figure 6.12a
Output as Printed by the Computer (Examples of Included Pages Follows in Figs. 6.12b and 6.12c)

commitment requires. The due date is compared to the present date, NOW, to establish the time remaining if the project is to be finished by the committed date, i.e., DUE-NOW 19.4 indicates there are 19.4 weeks remaining prior to the committed date.

Activities are indicated by numbers of the form 30–12, where the 30 would identify the organizational component responsible and the 12 would represent the specific activity. Next to each activity description is a list of the predecessor activities.

R. B. RICHARDS

PROJECT BEGINNING	1/08/79
NOW	3/19/79
DUE COMMITMENT	8/03/79
COMPUTED FINISH	10/10/79
DUE - FINISH	-9.4
DUE - NOW	19.4

ACTIVITY	DESCRIPTION	PRECEDENCES	DUR	EB
10-01	DEV MATERIAL LIST	0	0.4*	1/08/79*
10-02	DEACTIVATE OLD LINE	10-01	1.0	1/10/79
20-05	PROCURE PIPE	10-01	20.0	1/12/79*
30-03	ERECT SCAFFOLD	10-01	8.8	1/16/79*
30-12	REMOVE OLD P & V	10-02 30-03	8.4	1/18/79*
20-10	PROCURE VALVES	10-01	21.6	3/19/79
30-06	PREFAB PIPE SECTIONS	20-05	4.0	6/04/79
30-07	PLACE NEW PIPE	30-06 20-10 30-12	3.2	8/20/79
30-11	PLACE VALVES	30-12 20-10	0.8	8/20/79
30-08	WELD PIPE	30-11 30-07	0.8	9/12/79
30-09	FIT UP PIPE & VALVES	30-08	0.8	9/18/79
30-13	INSULATE PIPE	30-08	2.4	9/18/79
10-14	PRESSURE TEST	30-09	0.6	9/24/79
30-04	REMOVE SCAFFOLD	30-13 30-09	0.4	10/04/79
10-15	CLEANUP & STARTUP	30-04 10-14	0.4	10/08/79

Figure 6.12b
Project Summary and Detailed Schedule for All Activities Requested.

PAGE 1

LB	EF	LF	SLACK	ALLOC	TROUBLE	COMMENTS
2/27/79	1/10/79*	3/01/79	7.2	7.6		FINISHED
6/13/79	1/17/79*	6/20/79	21.8	23.6		FINISHED
3/01/79	6/04/79	7/23/79	6.8	28.4		BEGUN
4/18/79	3/19/79	6/20/79	13.2	24.4		NO FINISH GIVEN
6/20/79	3/19/79	8/20/79	21.6	33.2	1A	FINISH DELAYED
3/19/79	8/20/79	8/20/79	0.0	41.6		BEGINNING DELAYED
7/23/79	7/02/79	8/20/79	6.8	52.4		
8/20/79	9/12/79	9/12/79	0.0	66.4		
9/06/79	8/24/79	9/12/79	2.4	66.4		
9/12/79	9/18/79	9/18/79	0.0	70.4		
9/27/79	9/24/79	10/03/79	1.4	73.4		
9/18/79	10/04/79	10/04/79	0.0	73.6		
10/03/79	9/27/79	10/08/79	1.4	74.8		
10/04/79	10/08/79	10/08/79	0.0	76.4		
10/08/79	10/10/79	10/10/79	0.0	77.2		

R E T U R N   F O R M
MAIN OF CHEM PLANT
PERIOD   3/19/79 to   4/20/79

R. B. RICHARDS

TO BE RETURNED   4/25/79

ACTIVITY DESCRIPTION	PRECEDENCES	DURATION	
		OLD	NEW
30–03 ERECT SCAFFOLD	10–01	8.8	__.__
30–12 REMOVE OLD P & V	10–02 30–03	8.4	__.__

TROUBLE KEY
WILL BE UNABLE TO MEET
    1   SCHEDULE
    2   BUDGET
    3   SPECIFICATIONS
BECAUSE OF LACK OF
    A   QUALIFIED PERSONNEL
    B   FACILITIES
    C   MATERIALS
    D   DECISIONS
    E   EXPERIMENTAL OR TECHNICAL RESULTS
    F   NEEDED INFORMATION
    G   REALIZABLE SPECIFICATIONS

**Figure 6.I2c
Return Form**

The columns are labeled with abbreviations for DURation, Earliest Beginning, Latest Beginning, Earliest Finish, Latest Finish, SLACK, ALLOC, COMMENTS. ALLOC is $2 \times EB + DUR + SLACK$. This is used in allocating resources, as explained in section 6.13.

The time is given in weeks, based on a 5-day week, i.e., each day is 0.2 week. New Year's Day, Memorial Day, Fourth of July, Labor Day, Thanksgiving, and Christmas are treated as holidays.

Actual beginning or finish dates given in the input are indicated by an asterisk (*). If both beginning and finish dates are given, a FINISHED comment appears. If an actual beginning date but no finish date has been

PAGE 1

BEGINNING		TIME REMAINING	FINISH		TROUBLE
SCHEDULE	ACTUAL	TO FINISH	SCHEDULE	ACTUAL	
1/16/79*	__/__/__	__.__	3/19/79	__/__/__	1 2 3
					A B C D E F G
1/18/79*	__/__/__	__.__	3/19/79	__/__/__	1 2 3
					A B C D E F

CHECK HERE IF NO CHANGE __

IF BEGUN, "TIME REMAINING TO FINISH" MUST BE FILLED IN.

---

given, a BEGUN appears in the comments. If an activity has not been given a finish date in the input but a succeeding activity has begun, there is an inconsistency. This is indicated by the comment, NO FINISH GIVEN. If an activity's beginning is computed to be earlier than the date given in the summary as NOW but no actual beginning date has been given, the program assumes that the activity has not begun. Thus, it replaces the computed beginning time for this activity with date NOW and indicates BEGINNING DELAYED. Similarly, if no actual finish is given for an activity computed to finish prior to NOW, then the date NOW replaces the computed finish and FINISH DELAYED is indicated. These replacements also affect the computation of succeeding activities. If the beginning or finish of an activity has been delayed and the activity is critical, the activity should be given special attention.

We wish to make it easy for each of the managers in the system to report the status of his work to be used for updating input. If the manager must be involved in a great deal of red tape to make these reports, his reporting will be compromised. In standard reporting schemes, the manager usually begins with a standard reporting form, then fills in a great deal of descriptive information such as the date, his function, what project he is reporting on, the activities he is to report on, the address to which the form is to be returned, etc. In this system, the computer prints a return form "tailor made" to this manager, containing all of the known information, and leaving blanks for only the new information to be reported (Figure 6.12d).

Those activities for which the manager has reporting responsibility will appear on the return form. The manager fills in only those blanks that are pertinent. If, during the reporting period, an actual beginning or finish has occurred, he writes it in the appropriate space. If the manager has a new best estimate of duration, he enters that. If the network has changed to represent a different way of accomplishing the project, he may cross out or add predecessors, or add activities. The computer also addresses the return form so the manager can tear off the pages giving the form and its address, staple, and put it in the plant mail.

Of particular importance is trouble reporting. It is desirable to report troubles or anticipated troubles as soon as they become apparent. Troubles should be indicated and their resolution planned before their effect shows up as a slipped schedule. Thus, it is desirable to have the facility of reporting troubles on the return form. Management can then decide how the trouble should be resolved, and modify the schedule accordingly.

The trouble type is indicated by circling the appropriate keys on the return form (Figure 6.12d). This feature allows one to indicate troubles or anticipated troubles for early rectification before they appear as slippages in the schedule. The trouble indicator on the schedule listing (Figure 6.12b) may be controlled so as to appear only in the reports to individuals authorized to see troubles on the particular activity.

Another useful output is the Gantt chart, which shows a bar for each activity representing the interval of time from beginning to finish.

6.13 Resource Allocation

Neither PERT nor the previously described techniques handle, by themselves, the order in which to allocate limited men and machines to the activities. This is frequently referred to as the resource allocation problem.

The continuous resource allocation problem implies that any proportion of a resource can be assigned to an activity and that proportion can be changed anytime during the activity. Some man-scheduling problems fall into this category. For this case, the approach is to allocate more resources to critical activities and less to noncritical ones until they are all equally critical.

Discrete resource allocation implies that the resource in question cannot be proportioned, and once the activity is begun it continues until it is finished. The assignment of machines with a high setup cost fall into this category.

Many problems fall somewhere between the continuous and the discrete conditions; there is a nominal setup cost and/or only a few people or machines. We shall develop an index based upon solving the discrete problem. This index can be used with the slack and with the exercise of judgment to reasonably minimize the number and costs of resource changes for these intermediate problems.

The manager may add precedences to simulate his decisions on the order in which to allocate resources, i.e., if a resource is allocated to activity 1 before activity 2, activity 1 becomes a predecessor to activity 2. Seeing the consequences of his choice, the manager may try others until he is satisfied. The scheduling tool indicates to the manager the most likely good choices for these assignments.

It would appear that the calculation of earliest and latest beginning and finish times and slacks may be of assistance in making these resource allocation decisions. For example, let us consider that we have two activities which require the same machine. The earliest beginning times and slacks are first computed without restrictions on the availability of the required resources. (To illustrate the point, we draw the diagrams in PERT form where the length of the activity line represents its duration.) Consider Figure 6.13 and Table 6.9.

Activity 1 is ready to be started immediately, and is critical. If activity 2 is placed on the machine first, the machine will be idle 2 units of

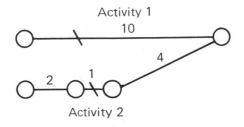

Figure 6.13
No Resource Constraint

TABLE 6.9: NO RESOURCE CONSTRAINT

Activity	Earliest Beginning Time	Duration	Slack
1	0	10	0
2	2	1	3

time waiting for a noncritical activity which is not ready to begin, while a critical activity is waiting to use the machine. It would then appear that activity 1 should be put on the machine first, as in Figure 6.14.

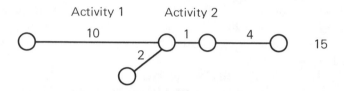

Figure 6.14
Activity 1 Before Activity 2

But this conclusion is wrong. If activity 1 is put on the machine first, the project is delayed 5 units of time. If, however, activity 2 is put on first, the project is delayed only by 3 units of time, as in Figure 6.15.

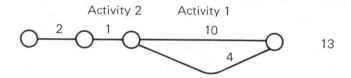

Figure 6.15
Activity 2 Before Activity 1

In this example it was better to put the activity which begins later and has more slack on the machine first because it ties up the machine for a shorter time. But we see that somehow the earliest beginning time, duration, and slack must all be balanced to make a good decision on loading resources. Let

t_{eb} = earliest beginning

t_{ef} = earliest finish

S = slack

D = duration

$t_{lf} = t_{ef} + S$ = latest finish

d_{12} = delay of project due to activity 1 using the machine before activity 2

d_{21} = delay of project due to activity 2 using the machine before activity 1

T = earliest finish time for project

$t' = T - t_{lf}$ = longest path required to do the activities following

Then

$$d_{12} = t_{ef1} - t_{eb2} - S_2$$
$$d_{21} = t_{ef2} - t_{eb1} - S_1$$

Let us go through how we obtained the above equation for d_{12}. Activity 1 is not delayed, but may cause activity 2 to be delayed. Activity 2 actually starts at 1's earliest finish time, whereas it might have started at its earliest beginning time. Thus, the delay to activity 2 is $t_{ef1} - t_{eb2}$. The delay to the project is the delay to the activity minus its slack. If d_{12} is zero or negative there is no delay to the project.

Consider $d_{12} - d_{21}$. The equivalent equations can be written in any of the following forms:

$$d_{12} - d_{21} = (t_{ef} + t_{eb} + S)_1 - (t_{ef} + t_{eb} + S)_2$$
$$= (2t_{eb} + D + S)_1 - (2t_{eb} + D + S)_2$$
$$= (t_{eb} + T - t')_1 - (t_{eb} + T - t')_2$$
$$= (t_{eb} + t_{lf})_1 - (t_{eb} + t_{lf})_2$$

The sign of this quantity indicates which delay is greater. This is then equivalent to comparing the size of the quantity $t_{eb} + T - t'$ for each activity. The activity with the lowest value of $t_{eb} + T - t'$ should be loaded on the machine first. This appears reasonable, since the early beginning time, early project completion time, and a long time required to do the activities following contribute to early loading.

If we are considering only the activities within one project, then T is the same for both activities being compared, and thus can be ignored. Then we compute $t_{eb} - t'$ and load first the activity with the lowest value. We then recompute t_{eb}, as it may have changed due to loading the previous activity, and apply the rule again. (Note that t_{eb} is the longest path from the beginning of the activity to the beginning of the project, and t' is the longest path from the end of the activity to the end of the project.)

The method described for resource allocation is not an algorithm guaranteed to give an optimum answer, but has been shown to give consistently good answers compared to other available methods.

A comparison was made between this method and a widely used method which considers only earliest beginning and slack. The project concerned the building of a highway bridge. There were 52 activities involving four types of resources. The results are shown in Table 6.10 as a function of the availability of each of the resources.

When the resources are unlimited, the two methods give identical results. Our method always does at least as well as, and usually better than, the other method. Note that in going from schedule 2 to schedule 3, the other method has the idiosyncrasy of computing a shorter schedule despite less resources.

TABLE 6.10: COMPARISON OF METHODS

	R_1	R_2	R_3	R_4	This Method	Other Method
(1)	Unlimited				217	217
(2)	25	25	11	7	224	231
(3)	25	25	10	7	224	224
(4)	25	20	20	5	273	280

(Resources)

6.14 Cost Planning

In planning a project we assign durations and predecessors, and then compute the resulting project completion date to see whether it is consistent with the required date. If it is not consistent, we may wish to expedite the schedule until the required finish date is met. The first activities to be expedited should be those on the critical path for which we can obtain the greatest time savings for the least cost.

Assume that for each activity we have a normal time and cost, and a crash time and cost. We assume a straight-line relation between them as in Figure 6.16. Let us call a_i the cost of expediting activity i by 1 unit of time. The flatter the line above, the lower will be a_i. We assume that once we get to the crash schedule, the cost of expediting each additional unit of time increases sufficiently to assume it would be uneconomic to expedite beyond this point. Thus, assuming just a straight-line relation between normal and crash is probably not too bad a compromise between accuracy, on the one hand, and the cost of data we must collect on each activity, on the other hand.

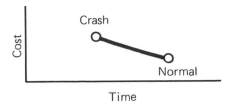

Figure 6.16
Activity Cost—Time

Now let us consider the whole project. We begin by setting every activity at its normal time and cost, and computing the time and cost for the project. We then apply the following rule: Expedite the activity on the critical path having the lowest value for a_i until either it goes to its crash point, or else some other path becomes critical. We do this again with the new schedule, etc.; by this means, we get a series of schedules. This basic rule is simple and generally works quite well. There are some unusual conditions, however, where the rule does not give quite the best answer. A method has been developed based upon parametric linear programming that guarantees the optimal solution under all circumstances. The mathematical details of this method can be found in the literature [Kelly:61].

Each schedule has a project time and cost which we plot on a time-cost curve for the whole project, as in Figure 6.17. For any given time, a point on this curve corresponds to the least cost schedule.

This curve represents direct costs associated with the activities. Generally, we also have indirect costs due to capital tie-up, overhead, schedule premiums, or penalties. The indirect costs generally increase with time. Thus, adding direct and indirect costs together will tend to give a U-curve, as in Figure 6.18.

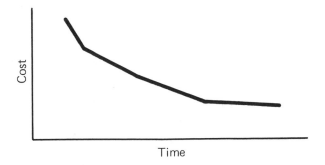

Figure 6.17
Project Direct Cost—Time

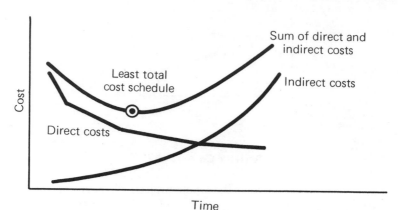

Figure 6.18
Project Total Cost—Time

From the sum of the direct and indirect costs, we can pick out a schedule which represents the least total cost.

This technique, variously referred to as the critical path method or critical path planning and scheduling, has been widely and successfully used, particularly in the construction industry.

6.15 Cost Control

As the project proceeds, management wishes to compare the progress with the plan to determine what activities are causing the greatest cost overrun, and what the new projection of the total project cost is. To meet these needs, let us consider a cost control technique as follows.

Several categories of cost will be established, for example, man-hours at union wages, man-hours exempt, purchased parts, facilities, etc. Costs are broken into categories on the basis that a projection made within one source of cost is more valid than projecting costs lumped together from several different types of sources.

To explain how we project the cost in each category, for each category let

E_t = originally estimated cost for total completion
E_p = originally estimated cost for that portion of the work for which costs have been reported
A_p = actual cost for partial completion as reported (i.e., A_p and E_p represent the actual and estimated costs for the same work)
E_t' = computed re-estimate for total completion

Toward the beginning of the project we have very little information upon which to base a trend we can apply to the remaining costs. Thus, we are inclined to add the originally estimated costs for the remainder of the project to the accrued costs: $E_t' = A_p + (E_t - E_p)$.

Toward the end of the project, when we have enough information to establish a trend, it would appear appropriate to scale the remaining costs by A_p/E_p:

$$E_t' = A_p + \frac{A_p}{E_p}(E_t - E_p)$$

The following formula reduces to the above formulas at the beginning and end of the schedule. Thus, in general, we use the following procedure:

$$E'_t = A_p + W \cdot (E_t - E_p)$$

where

$$W = \frac{(E_t - E_p)}{E_t} + \frac{A_p}{E_p}\frac{E_p}{E_t}$$

To this, we must add one further feature. Some deviations from planned costs may be known to be unique and not have any effect on further costs in its category. Such costs should be exempted from the above projection.

This method of cost following and projection offers an alternative to the classic scheme of plotting accrued costs versus time.* It is normally assumed that if the actual accrued cost at a given time is less then originally estimated for that time, then the final cost can also be expected to be lower than the estimate. But quite the contrary may be true. Some costs accrue as a function of the work accomplished (e.g., material costs), while others accrue as a function of the time transpired (e.g., salaries, rents, capital use costs). Thus, if the actual accrued cost to date is less than estimated because less has been accomplished at this time, there is good reason to believe that final costs may run higher, not lower, due to the longer project time.

We are interested in directing management's attention first to the activities and categories that have the greatest likelihood of requiring management action. For this purpose, we propose the use of the concept of

*The classic curve of accrued cost versus time follows the same shape as the curve showing the population of fruit flies multiplying in a confined space. It grows rapidly at the beginning, and levels off at the end. It is interesting to conjecture about the analogy between the manning of projects and the population of fruit flies.

manageable anticipated overrun. The anticipated overrun is defined as $E' - E_t$, in other words, the best re-estimate of the total costs minus the originally estimated or budgeted cost. Management action can do very little about costs which have already been overrun. Therefore, we multiply the anticipated overrun by the proportion left to complete to indicate how much of the anticipated overrun may still be manageable. Manageable anticipated overrun is not intended as a rigorous control quantity, but only as an indicator as to which items management's attention should be directed first.

$$\text{M.A.O.} = (E'_t - E_t)\frac{(E_t - E_p)}{E_t}$$

Valid objections have been raised with respect to estimating the completion of activities in progress for input to such a scheme. The best alternative appears to be to estimate the remaining costs. The preceding techniques, however, may still be used. Note that the error in estimating completion or remaining costs affects only those activities in process. Presumably, these errors will be diminished in their effect on the estimates of total costs if only a small fraction of the total number of activities are in process at any one time. The remaining activities are either finished or not started.

6.16 Scheduling Production of Multiple Units—Queuing

Let us consider the extension of scheduling techniques to handle production of a number of units where each unit goes through the same operations with the same precedences and durations. Note the example in Figure 6.19, where the numbers below the circles are durations for each operation on each unit.

Here the sequence of operations 1–5 is critical until the first unit has passed through all five operations. Thereafter, a unit is produced by the sequence 1–5 every four time periods. Production is thereafter limited by operation 10, which produces one unit only every five periods of time. Thus, we see that the critical operations can change with the number of units that have been processed. Certainly, we could handle this by considering each operation on each unit as a separate activity. However, with a large number of operations and units, this may involve the

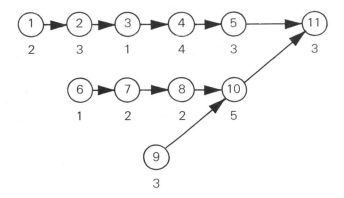

Figure 6.19
Precedence Network Diagram

computation of a large volume of numbers in which trends may be obscured.

Let us make a few assumptions. The reasoning for these assumptions will become self-evident by going through our example. Let us represent a time (e.g., an earliest or latest beginning or finish time) as follows: $t = a + b \times n$, where t is the time for the nth unit. Now we define an "addition" and "subtraction" of a duration as follows:

$(a + b \times n)$ plus a duration d is equal to $(a + c_s) + c_l \times n$

$(a + b \times n)$ minus a duration d is equal to $(a - d) + b \times n$

where c_s is the smaller of b and d, and c_l is the larger of b and d. The result is again of the same form as $a + b \times n$.

Let us assume we have an activity with two predecessors. The earliest beginning would have to wait for the larger of the two finish times of the predecessors, which are expressed as $a_1 + b_1 \times n$ and $a_2 + b_2 \times n$. The latest time for any n is the larger of these two expressions at that value of n. If either $a_1 > a_2$ and $b_1 < b_2$, or $a_1 < a_2$ and $b_1 > b_2$ (i.e., the relation between the a's is the opposite of the relation between the b's), then one of these expressions will be greater for low values of n and the other will be greater for large values of n. We must carry both expressions, and consider the larger of the values given by the two lines. Similarly, when we work with successors and latest times, if the relation between the a's again is opposite to the relation between the b's, we carry both expressions, but this time we consider for each n the smaller of the values given by the two lines.

Using these new rules for the meaning of "addition," "subtraction," "larger," and "smaller," we may proceed to compute earliest and latest beginning and finish times very much as we did earlier for simple schedules. However, we also have one further condition to consider in computing earliest beginning times. Not only must an operation on a unit wait for the predecessors to be finished, but it must also wait for the previous units to finish going through this operation. This is taken care of by the following device. We carry in our table a latest predecessor column and "add" our duration to this to get the earliest finish time. Then we "subtract" the duration to get the proper earliest beginning time.

This method of calculation is illustrated by carrying out the example of Figure 6.19 in Table 6.11.

TABLE 6.11: Schedule Calculation for Multiple Units

Activity	Duration	Last Predecessor		EF		EB		LB		LF	
1	2	0	0	0	2	-2	2	-8	5	-6	5
								-4	4	-2	4
2	3	0	2	2	3	-1	3	-6	5	-3	5
								-2	4	1	4
3	1	2	3	3	3	2	3	-3	5	-2	5
								1	4	2	4
4	4	3	3	6	4	2	4	-2	5	2	5
								2	4	6	4
5	3	6	4	9	4	6	4	2	5	5	5
								6	4	9	4
6	1	0	0	0	1	-1	1	-5	5	-4	5
								-1	4	0	4
7	2	0	1	1	2	-1	2	-4	5	-2	5
								0	4	2	4
8	2	1	2	3	2	1	2	-2	5	0	5
								2	4	4	4
9	3	0	0	0	3	-3	3	-3	5	0	5
								1	4	4	4
10	5	3	2	5	5	0	5	0	5	5	5
								4	4	9	4
11	3	5	5	8	5	5	5	5	5	8	5
		9	4	12	4	9	4	9	4	12	4

The earliest and latest beginning and finish times for operation 10 and the earliest finish times for its predecessors are shown in Figure 6.20. The slack is the shaded area between the earliest and latest beginning or finish lines. Initially, the sequence 1–5 and 11 is a critical path. Operation 10 begins with a slack of 3, but becomes critical at $n = 4$ and remains critical thereafter. Note that although operation 10 is then critical, there is no critical path in front of operation 10. In fact, its predecessors' activities can supply parts faster than operation 10 can use them.

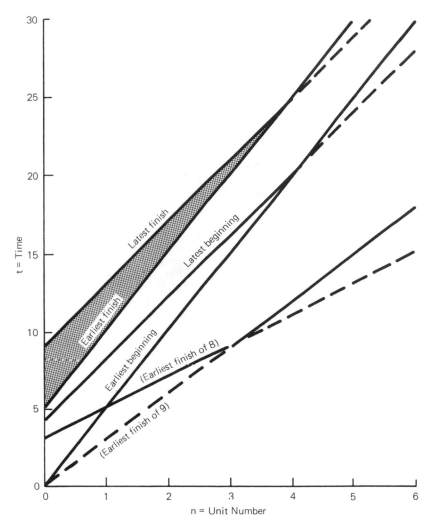

Figure 6.20
Schedule Times for Operation 10

Let us assume that every operation on every unit starts at its earliest beginning time. Then we can use the graph in Figure 6.20 to tell what the queue (or accumulation of unprocessed parts) in front of operation 10 is. For example, if operation 9 is supplying parts at its earliest finish time while operation 10 uses them at its earliest beginning time, then the difference is the accumulation. At time $n = 13$, operation 9 has produced four parts while operation 10 has used two. If we wish to start operations at their latest beginning times, or somewhere between the earliest and latest, we can redistribute the inventory of accumulated, unprocessed parts.

7 The Engineering Design of Systems

7.1 Management Models—Who's Kidding Whom?

This chapter shows how the concepts of critical path scheduling, which have proved so successful in the construction industry, have to be extended to handle circuits before they can be applied with equal effectiveness to engineering. The techniques developed earlier in this book make that extension possible.

But first we must investigate some of the problems inherent in the use of currently available engineering management systems before we can appreciate the simplicity and power of the techniques presented here.

Many management systems for planning, scheduling, and controlling the engineering design of systems have acquired a bad reputation. Let us discuss some of the reasons.

The reporting of progress and the management decisions based on that progress must assume some model. That model may or may not be explicitly stated. For example, the model may be a file of independent schedule dates which assumes that every task can be done independently of every other task. Or the model may be a critical path diagram showing which tasks depend on which other preceding tasks.

Often the model is not a valid representation of how the work is or could possibly be done. For example, the file of independent dates may ignore the fact that certain tasks cannot be started until other tasks are completed. Or a critical path schedule may ignore the circuits which are so characteristic of engineering. Many models, even though valid when the work begins, are not sufficiently versatile to accommodate unexpected developments which occur as the project unfolds. Updating the times and changing the precedence relations in the schedule may be too difficult to administer and thus cause the schedule not to be kept up-to-date. At best, such models may represent a fiction that does not reflect the actual state of affairs. This can lead to erroneous management decisions. At worst, the engineers accept the constraints of a bad model, the model becomes a straitjacket, and little or no effective work gets done. The engineers often find themselves faulted for lack of discipline because of their unwillingness to conform to what they recognize as an untenable scheduling model.

Models and schedules are often set up by a small group of overlords who have neither full knowledge of all the details of the project nor seek the consultation of those who do. This schedule is then imposed upon those who do the work. These people, who do have the knowledge of the details and problems, may have good reason to disrespect the fictional basis of the schedule.

It is quite common to find there is one group that does the work by whatever means it can, while another group maintains the fictions used by management to make their decisions.

Often there is an individual on the project who has in his head a good model of the design plan. This may be quite sufficient, if the project is small enough that he alone can make all the detailed decisions which depend upon that plan. If, however, the project is large enough so that some or all such detailed decisions must be delegated, then it is necessary that the model or plan be so stated that it can be understood in the same way by all those who must make the delegated decisions. Otherwise, individual decisions may be made at cross-purporscs.

It is not unusual for engineering management to bemoan that their engineers are an undisciplined lot of technically brilliant prima donnas. But these same managers place the burden for the discipline on the engineers instead of providing a tool to help them, saving the engineers' time to do engineering. Let us illustrate what can happen.

Each engineer is asked to give estimated completion dates for his tasks without being told when the tasks he depends upon are scheduled to be finished. Thus, it becomes each engineer's responsibility to solicit from fellow engineers their estimated finish dates. But these fellow engineers may not have traced the schedule dates of their predecessors and estimated their own finish dates yet. This is like having a group of people, each with one piece of a jigsaw puzzle, all running around trying to fit their piece into the whole puzzle; but the part they have to fit into may not have been placed in position yet.

With such a system there certainly should be no mystery as to why so much engineering time is lost thrashing around for information which is not ready yet. Some may ask why the process appears to be so undisciplined (it is), why so many inconsistencies show up in the schedule such as a task scheduled to be completed before tasks it requires are scheduled to be started, and why the schedule is so often not met. How can you expect anyone to meet a schedule when you provide on a silver platter the ready-made excuse that the schedule is inconsistent and thus impossible?

The integration of schedules by thrashing can easily lead to a particularly vicious circle. Let us say that 10% of an average engineer's time is taken up with this thrashing process. The work load starts to increase and schedules begin to slip. Management, without a good model of cause and effect in the schedule process, cannot trace the real cause of the slippage, so they increase the pressure on the individual engineers to meet the schedule. The response is for the thrashing to go up to 20%. The resulting work capacity decreases. The schedule slips further. Everyone becomes angry. The real problems are still not understood, the thrashing increases, etc. Anyone who has had much experience in scheduling under pressure will probably not find this hard to believe.

Each engineer sees only a small view of the overall scheduling puzzle. How much more civilized it would be to ask each engineer to submit his piece of the puzzle, i.e., what tasks are required as predecessors and how much time is required to perform the task, to a central facility where all of the pieces can be put together in an orderly way to generate a consistent schedule. There is no reason why this central facility should not be a computer. This process of putting the pieces together can easily be programmed. However, the resulting proposed schedule should then be subjected to careful scrutiny and revision.

The break up of the project into tasks and the assignment of the tasks to individuals is done from the top down. Then the model we consider here develops the preliminary schedule from the bottom up. The basic definition of the project in terms of precedence relations, time, and resource data are defined by those persons most directly responsible for the individual tasks and thus most intimately aware of the real details and problems; a maxim of scheduling says that the schedule tends to become more accurate as more detail is considered. Their data and the resulting schedule should be reviewed and negotiated with management. The analysis of the precedence matrix may be done in conjunction with management. But those who work is being scheduled should be involved and jointly responsible with management for the realism of the schedule which will discipline their work. This leads to better acceptance as well as added realism.

Also, the individual engineer knows that a good schedule model works for him too. It not only shows where his problems are, but it honestly directs management to where the problems are. Thus the engineer is held responsible for that which he can control, but not for that which is within someone else's control.

The design structure matrix discussed below presents a clear picture of the design plan which can be understood in the same way by everyone responsible for making decisions. This should be of great help in obtaining detailed decisions which are compatible with the overall plan.

If a critical path procedure were used without the guidance of a precedence matrix, it would be extremely difficult to operate a bottom-up system, for critical path scheduling used in the conventional way requires the maintenance of the fiction that there are no circuits in the network. But in fact, as we shall see in the following section, there are circuits, and these circuits are revealed in the bottom-up approach.

The model we shall discuss shows clearly the information flow which occurs in the design process; further, when the results of any task are changed it shows what other tasks are affected. It has been the author's observation that some of the most serious problems in the engineering process arise not because of a lack of technical competence, but because of deficiencies in information transfer. Very often design errors, added cost, and lost time occur because someone whose work is affected has not been made aware of the change. One group of engineers expends great effort optimizing, unaware of a change which makes their part of the design incompatible with the rest of the system.

7.2 The Role of Engineering: To Resolve Circuits

The engineering work required to design a system can be broken into a number of tasks. Examples are:

1. Determining a certain piece of the data which makes up the description of the system, how it is made, how it behaves, or how it is to be operated
2. Making a decision
3. Running a computer program or performing a design procedure
4. Producing a document or drawing that records certain data
5. Reviewing the design of some aspect of the system

These tasks will be constrained by a set of precedence relations showing which tasks need the results of which other tasks. These constraints may be technical. For example, the determination of a temperature T may require a prior determination of a heat flow Q. Or the constraints may be nontechnical, such as a review cycle that cannot be started until the requisite document is drafted.

In engineering it is quite likely that the technical precedence relations will involve circuits. For example, the temperature T depends upon the heat flow Q, Q depends upon the heat transfer coefficient K, but K depends upon the temperature T, thus completing the circuit. One may begin by assuming an estimate for one of these values in the circuit, say, K. This K may be used to compute Q, which is used to compute T, which is used to compute a new value of K. If the new value does not agree satisfactorily with the estimate, we can use it to repeat the process until a satisfactorily consistent set of K, Q, and T is obtained. This is, of course, the process of iteration.

There may be an iteration involving only a small set of tasks. Or an iteration may involve the whole system, such as a heat exchanger, leading to a whole redesign.

We have made the point that in engineering the precedence relations involve circuits. The role of engineering is to resolve these circuits so that no circuits occur during fabrication. Engineers can resolve circuits using pencil and paper, computers, or perhaps tests of pilot models at much less cost than these circuits might be resolved by trial and error during fabrication. For example, consider the cost of installing and welding into place a hundred thousand dollar heat exchanger and then discovering that it is too small to do the job, then ripping it out only to find there is not room enough to install another heat exchanger of the right size, etc. Of course, this would not occur because the right size heat exchanger would have been selected in the first place with engineering tables and a few dollars worth of calculations on a computer.

Critical path techniques have been successfully used for planning, scheduling, and controlling the fabrication of systems. But standard critical path techniques make no provision for handling circuits. As we have seen, fabrication does not involve circuits because the circuits are resolved by engineering. But to plan, schedule, and control engineering we do need a tool which can deal explicitly with circuits.

7.3 The Engineering Design Plan

Let us assume that we have defined the tasks required to produce a complete design of the system and have determined for each task what its predecessor tasks are. We can display the tasks and the precedence relations between tasks with a precedence matrix. Each row and its corresponding column represent a task. If task A requires the results or estimates for the results of task B, then there is a mark in row A, column B. We saw an example of such a matrix for the design of an electric car in section 2.5 and Figure 2.4. The matrix, after being reordered by partitioning and tearing, is called a *design structure matrix*. The design structure matrix shows the engineering design plan (see Figures 2.7 and 7.15i).

Let us see how the design structure matrix shows the engineering design plan. The marks above the diagonal show where assumptions must be used to start the design iteration process. These assumptions are usually based upon prior experience with similar systems. The responsibility for these assumptions will thus fall upon the more experienced engineers. Each assumption which is made should be recorded. When the last task in the block is completed, a design review should establish whether the assumptions are valid, or whether another iteration of some or all of the tasks in the block is required. If another iteration is made the schedule should be modified accordingly. The marks below the diagonal show when certain tasks have been completed and what other tasks are ready to be started. The preparation of input corresponding to these marks is usually straightforward and can be delegated to junior engineers. Thus we sometimes say that the marks above the diagonal are the province of the overlings, while the marks below the diagonal belong to the underlings.

Sometimes it is necessary to shorten a schedule by beginning tasks before their predecessors are finished. The cost of beginning such tasks early is the necessity of making additional assumptions. These assumptions also must be recorded and later verified.

Thus, the design structure matrix is a plan showing where assumptions are made, the order in which the design tasks are done, where design reviews are required, and what needs to be verified. Using the design structure matrix for reference, each person in the design process has the same understanding of the design procedure. It can be very important to have everyone working according to the same plan.

7.4 Critical Path Scheduling—After Unwrapping the Circuits

Clearly, each circuit or block is iterated only a finite number of times; therefore, it is possible to unwrap each circuit by laying it out end to end the number of times it is to be iterated. One then has a graph without circuits. Once one has made an estimate of how many times each block is to be laid out and has made an estimate of how long it will take to perform each task for each iteration, then classic critical path techniques can be applied to obtain a schedule. Thus the design structure matrix does not represent a technique which replaces critical path techniques for scheduling. Rather, it is a tool to help set up and operate a critical path schedule where there are circuits, as in engineering.

How many times a block is iterated will depend upon how good are the original estimates used for the marks above the diagonal, how sensitive the other tasks are to the errors in these estimates (and thus whether, and how quickly, the iteration converges), and how refined the data has to be for the design. Usually engineers want to refine the design by doing more iterations, while management prefer to avoid the costs in time and money of doing the added iterations.

It is not usually known for sure at the time the original plans and schedules are drawn up how many times each block must be iterated or how long it will take to perform each task in each iteration. However, it usually is necessary to make some best possible guesses to estimate the time and resources which will be consumed by the project. As the project proceeds one can replace estimates with known values for work accomplished to date and refine estimates and factor in new decisions to reschedule the work which lies ahead.

The design structure matrix and critical path scheduling have their greatest value not so much in the original planning and scheduling phase as in the continual replanning and rescheduling which one expects to occur as the project unfolds and unexpected situations arise.

A design review is required at the end of each major block. Part of the plan should be the decisions of what constitutes major blocks which require review and at what management level the review should occur. A design review may result in a decision to go on without further iterations, or a decision may be made to iterate all or part of the block. The decision whether to make another iteration may depend upon comparing the value of the improvement of the design to be expected from another iteration against the consequences to the schedule and engineering cost. At this time trial critical path schedules with and without another iteration might be prepared and compared. Once the decision is made it must be reflected in the working schedule.

In estimating the duration of each iteration of each task, we must note the following: Often there will be a setup time required only the first time the task is performed. Each time the task is done thereafter, due to learning, the duration may be reduced by some percentage of the difference between the duration for the first time (excluding setup) and the asymptotic duration, i.e., the minimum time one can achieve after many iterations. Thus, the time for each iteration might be estimated by a formula such as the following:

$$d_n = d_\infty + (d_f - d_\infty) * l^{n-1} + s \text{ if } n = 1$$

where d_n = the duration for the nth iteration

d_∞ = *asymptotic duration, i.e., the minimum duration after many iterations*

d_f = duration excluding setup time for the first iteration

l = learning factor, e.g., $l = 1$ implies no learning

s = setup time for the first iteration only

n = iteration number

Consider the graph and precedence matrix shown in Figure 7.1. Here there is an outer block involving six tasks; within that outer block there is an inner block involving three tasks. Assume that we figure we must iterate the outer block twice, i.e., a preliminary and a final design. Assume further that during the first iteration of the outer block we plan that the inner block will be iterated twice, while during the second iteration of the outer block the tasks in the inner block need be done only once. Then we can unwrap these blocks and lay them out as a critical path network; see Figure 7.2.

The critical path format used here is not the old PERT format which puts the activities on the lines. The precedence format used here puts the activities on the vertices and uses the lines to show the precedences. This corresponds to the way we have drawn graphs for tasks elsewhere in the book. Of course, this graph no longer contains circuits. Each activity in the critical path schedule represents a specific task performed during a specific iteration.

In Figure 7.2 the legend shows the interpretation of the numbers around the circles. EB and EF stand for the earliest beginning and earliest finish, and LB and LF stand for the latest beginning and latest finish. The project begins at week 5 and is required to finish at week 35. The rules for computing these times and slacks are given in chapter 6.

We have made the assumption that a task can be started only after each of the predecessors has been entirely finished. Sometimes it is useful to have a model in which one can specify for each predecessor what fraction of that predecessor must be complete before the task can begin, or allow a predecessor to be concurrent with up to some fraction of the task it precedes; see Figure 7.3. Of course the same effect can always be obtained by breaking the tasks into smaller tasks, but sometimes it is more convenient not to have to do this.

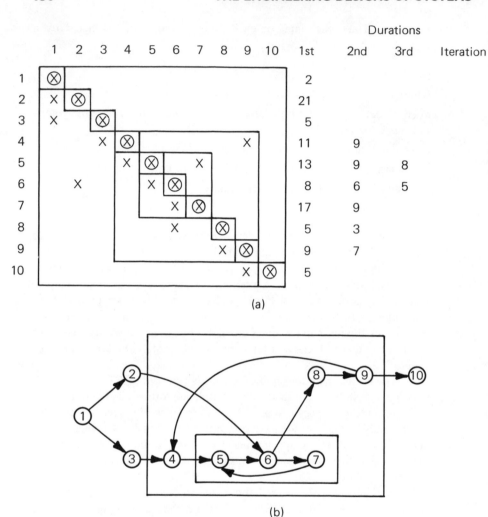

Figure 7.1
Precedence Matrix and Graph Corresponding to Critical Path Schedule in Fig. 7.2

If we record when each task is finished, then when any task is not finished as scheduled we can trace the cause of the hold-up with the help of a critical path graph or a precedence matrix. Where we traced effects by moving down columns to find successors, we can similarly trace causes by tracing across rows to find predecessors.

An active task is an unfinished task with no unfinished predecessors.

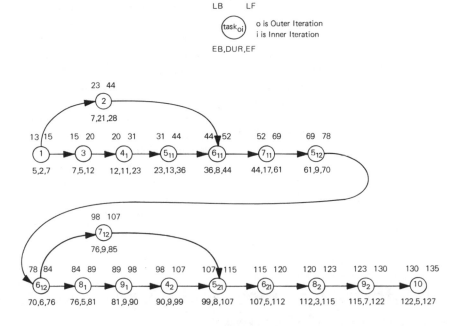

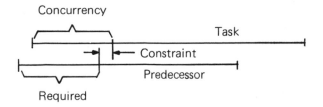

Figure 7.2
Critical Path Schedule Corresponding to Precedence Matrix and Graph in Fig. 7.1

Figure 7.3
Overlap between a Task and Its Predecessor

A manager has a chance of affecting the performance of an active task, but it is no longer possible to affect the performances of tasks already finished. He can affect those tasks not yet ready to begin to the extent of replanning, rescheduling, and reallocation of resources. A manager would particularly want to keep track of any schedule slippage of active tasks so that he can begin immediately to plan remedial action.

If a manager is concerned about the cause of the delay of any unfinished tasks, he can trace its predecessors back to the active tasks which affect it. Then he can determine whether the slippage can be remedied by applying attention and resources to these active tasks, or whether the slippage is already determined by completed tasks. In the latter case there may be a need to replan unfinished tasks and reallocate resources to make up the lost time.

When a task becomes active it does not mean that the work on the task has necessarily begun. The beginning of the work may be awaiting the availability of the engineer's time or of some other resource. An engineer working on other projects can lose track of the fact that a task on this project has become active. Thus the system should report to him when a task of his becomes active and what the urgency of that task is with respect to other active tasks requiring his attention on this or other projects.

7.5 Information Flow and Propagation of the Effects of Changes

The precedence constraints which occur in scheduling the fabrication of systems arise primarily because of the flow of materials or the physical conditions which must exist before the next step can begin. For example, concrete may not be poured until it is delivered at the place it is to be poured (i.e., a flow of materials) and the forms have been laid (i.e., an existing physical condition).

In engineering, the role of constraints due to flow of materials is replaced by constraints due to flow of information, e.g., the mounting can be designed once it is known what load it will have to take. We may still have constraints due to existing physical conditions, e.g., a report is verified only after it is finished. In either case the graph showing the precedence constraints can be used to show a propagation of the effects of a change both on the system design and on the schedule.

Assume that some portion or all of a system has been designed and now there is some change to be made. Such changes can arise from many sources:

1. Perhaps an error has been recognized either during the design or when a subsystem or the whole system is tested.
2. An opportunity for improvement in the design may be perceived.
3. The customer or the marketing department may propose changes in the specifications.

4. A review at the end of a block may show where assumptions have to be revised, requiring changes to tasks in the block.
5. In many businesses it is the practice to sell standard designs modified to meet specific customer requirements, in which case the specific requirements can be considered as changes to the standard design.

Thus it is common that one is concerned with changes which might affect only a small subset of all the tasks in the system. The design structure matrix can be used, as we shall see below, to determine this subset of affected tasks.

If only a small number of tasks is changed in a large system which has already been designed, the results of the predecessors to the tasks changed will already be known from the original design. This decreases the likelihood that the change will cause circuits. However, circuits can still occur.

If the system has not been completely designed at the time the change is introduced, many of the affected tasks may not have been done the first time. Thus it may be possible to incorporate the changes into these tasks with little or no added cost.

The design structure matrix will show the propagation of effects which occur as a consequence of a change occurring anywhere in the design process. Where going across a row of the matrix shows what precedes that task, going down a column shows what succeeds that task, i.e., what other tasks use the results of this task. When the results of a task change, one can go down the column corresponding to that task to see what other tasks are directly affected and thus subject to change. Then going down the columns associated with these affected tasks shows what tasks they affect in turn, etc. This process may be continued until one finds no new tasks with successors, or until the effects on the successors which remain are considered to be of no consequence.

Changes cause changes, in both hardware and documents. The documents changed are those that carry the changed parameters through the redesign process, and those that reflect the changes in the final system and/or hardware being designed. The maintenance of documents during the change process is often referred to as *design change control*. The design structure matrix can be an important part of the implementation of design change control. It shows where each change may cause other changes which must be reflected in updated documents.

The verification of the design presents a similar problem. Verification concerns the cross-checking of documents and decisions to ensure that the design has been done correctly and that all necessary design requirements have been met. Design requirements may arise from marketing or management considerations, the customer's specifications, legal and safety requirements, etc. The verification of the whole design implies that each document has been verified subject to the assumed validity of each of its inputs, and that all the specified revision levels of the inputs assumed in the verification have similarly been verified. If a change is made in a document after that document has already been used for the verification of another document, then that verification must be reconsidered. If, as a consequence of this reconsideration, the verified document were to change, then a new revision level must be issued and all other documents it affects must be reconsidered. The design structure matrix can be an effective tool in this design verification process.

It is a straightforward but tedious chore to trace these effects through a matrix of fifty or a hundred tasks. But it becomes completely impractical to trace the effects in a matrix of several hundred or thousands of tasks. Thus a time-share program, called ASPECT (Analysis of Structure and Propagation of Engineering Consequences Throughout), allows a user at a time-share terminal to type a list of the tasks he proposes to see changed. Then he may trace the consequences through the system task by task. When the user encounters a task he believes is not significantly affected and is thus not worthy of being changed, he may reject that task from the list so its consequences are traced no further. This task rejection feature is what makes it desirable to run such a program on an interactive (time share) computing system rather than on a batch system. See Figure 7.17.

There are several other useful features required of a program such as ASPECT. Often the same task is encountered more than once because it is the successor of more than one earlier task. Whenever such a task is repeated, the program should note this fact but not trace the successors of this task more than once.

When all the potentially affected tasks have been traced, ASPECT retains a list of them. ASPECT can then be used to retrieve various files of information associated with these tasks. These files contain information such as estimates for costs and for the parameters from which a computer program computes the duration for each iteration of each task, as in the formula discussed in section 7.4. The precedence information for the affected subset of tasks can be extracted from the original design structure matrix and analyzed to obtain a design structure matrix for just this subset of tasks. Once the user makes judgments on how many times each block is to be iterated, preliminary input to critical path and cost analysis programs can be generated by the computer. This input should be reviewed and modified to incorporate any further judgments not directly implied by the information in these files. This procedure is used to develop a quick estimate of the impact of the proposed change on the engineering cost and schedule for the system.

Once an estimated cost and schedule for the proposed changes have been arrived at, decisions can be made whether the changes should be done. Engineers frequently like to put in a number of changes to tune up a system to the last decimal place; marketing is inclined to propose to customers all sorts of changes which will enhance the likelihood of a sale. These changes are often proposed without a full appreciation of the string of consequences and their resulting costs and schedule delays. The design structure matrix with appropriate files plus ASPECT and a critical path program give management help in evaluating the merits versus costs of such proposed changes.

Up to this time we have considered changes which affect the data or the results of tasks, or changes which affect the number of iterations of a block or the time required to perform tasks. But we have not considered that the tasks themselves could change; for example, that a change could be made in the technology to be used in some part of the system.

Clearly, when the tasks themselves change, the matrix, or that portion of the matrix concerning the tasks affected, must be changed. For example, we may make a choice between one of two alternative processes for removing some pollutant. One process may be strictly mechanical while another may involve chemical precipitation. These two processes would require quite different types of equipment and would involve different design variables and design tasks to design them.

One may make up precedence data for each of the alternative technologies being considered. Matrices and estimated schedules and costs can be developed for each such technology using this data. These estimates may be considered along with the technical evaluation in deciding which technology to use. Once the technology is chosen, the precedence data for that technology is put into the system and used for the remaining planning, scheduling, and control of the project.

7.6 Information Flow and the Organizational Structure

The general principles applied to relate information flow to organizational structure are as follows: (1) one-way information flow across organizational boundaries is usually more formal and more difficult than information flow within an organization; (2) two-way information flow across organizational boundaries brought about by iteration is even more difficult.

Generally, communication between two individuals becomes more formal and more difficult as you have to go higher in the organization to find the lowest level manager common to the organization tree of both individuals. It is desirable to confine the responsibilities for the tasks within a block to the smallest level organizational unit capable of managing the number of people required to perform the tasks.

When an iteration occurs between tasks in different organizational components, there often occurs duplication of effort; each component feels some responsibility to check the validity of the information which comes from the other component. Some duplication of effort can be saved by minimizing the information which must cross such an interorganizational boundary.

It is often good strategy (see section 7.9) to tear and order the tasks in the matrix such that the blocks break down with smaller blocks within larger ones. For example, in Figure 2.7 we have a block of size 8 containing a block of size 6, which in turn contains a block of size 2. Where there are small blocks, the persons associated with the tasks within the block should be closely associated so they can communicate orally without having to clear their communication through some higher level of management. The results need not be documented until their iterations are complete.

But iterations through larger blocks involve more tasks and more people. Thus, communication will require more formal documentation. Where possible, it is desirable to structure the organization such that the communications within a block do not cross difficult boundaries between organizational units.

A design structure matrix may be set up to include tasks from outside the company. If circuits cross company boundaries some consideration should be given to changing the product scope so that the circuit is confined to within the company.

The most efficient organization for a particular project should be a reflection of the structure of the information flow inherent in the structure of the design process itself. This structure of the information flow is shown by the design structure matrix.

However, it often occurs that the organization has not been created to respond to the structure of the system being designed. One must then deal with the organization as he finds it. The insight into the information flow inherent in the system can be invaluable either for changing the organization, or in getting the job done despite the organization as it exists.

7.7 Information Flow and Document Definition

A document is a collection of data published at one time. How do we define what data should appear with what other data on the same document? There are two considerations in approaching this problem.

First, from the point of view of information retrieval, we would like to see closely related data on the same document. For example, if we obtain a document on the heat balance we will probably want to see a complete story, or at least a good summary, assembled on one document rather than having to retrieve a dozen documents with a little piece of information on each.

A useful device for defining and retrieving the documents describing the hardware of a system is a document tree. It is a form of a graph where each vertex represents a document. Each document describes or specifies a part or an assembly of parts in a parts breakdown. Figure 7.4 shows an example of a parts breakdown.

We mention the document tree in order to make the following distinction: A document tree is a graph which corresponds to how the hardware of the system is assembled; but the precedence matrix discussed here corresponds to a graph showing the information flow which occurs

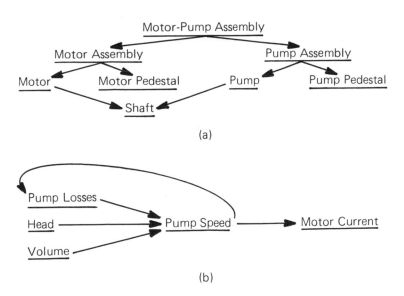

(a)

(b)

Figure 7.4
(a) Hardware parts breakdown/document tree compared to (b) information flow graph

during the design process. Note that the flow of information in the design process may cut across different parts and assemblies in the system. For example, the knowledge of a pressure may be required to design the pump, a pressure vessel, and a control system. These components would probably not be related in a common low-level assembly. Systems documents may be developed to represent various types of flows, such as a process diagram showing both the pump and the control system. The document tree corresponds to the parts explosions of the hardware and the systems documents showing the various process flows, etc. Transcending this we have an information flow which says which of these documents we must know about to produce certain other documents (see Figure 7.4).

Second, we shall consider the structure of the data on a document from the point of view of information flow. This we can do most easily by first asking how the information flow can be hampered by faulty definition of documents. Let us begin by describing some of the problems of document definition.

The first problem is the introduction of document circuits. A *document circuit* is a circuit which would not exist except for the definition of the documents; Figure 7.5 illustrates this. Document A contains two data sets, A_1 and A_2. Data set A_1 is a predecessor to a data set on document B, and some data set on document B is a predecessor to data set A_2. Thus we have a circuit between document A and document B even though there

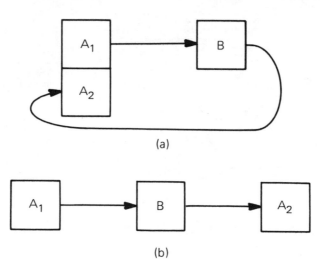

(a)

(b)

Figure 7.5
Unnecessary Circuit Due to Document Definition. (a) Document circuit between documents A and B. Data sets A_1 and A_2 on same document. (b) Dividing document A into data sets A_1 and A_2 on separate documents removes circuit.

is no circuit inherent between the data sets involved. This circuit can be torn by breaking document A into two documents, or putting either data set A_1 or A_2 on document B rather than on document A. These problems can be recognized by breaking the documents into the data sets they contain and making up the matrices which represent the precedence relations between these data sets. We usually recognize a document circuit if we see that circuits in the document-to-document matrix disappear when we go to a data set-to-data set matrix.

The second problem arises when one uses too coarse a definition for the tasks. Then it is possible to get too many successors when tracing the propagation of effect; see Figure 7.6. Assume that the effect of a change is traced to document A. A is succeeded by B, and B is succeeded by C. Thus, if one were using just the document-to-document relations, one would conclude that document C is also a candidate for change. If, however, we considered the data set-to-data set relations, we would see that no data set on document C would be affected. Thus, by breaking the documents or tasks into further detail, we do not pick up as many extraneous documents or tasks when tracing effects.

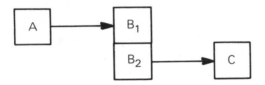

Figure 7.6
Trace Extra Predecessors Due to Document Definition. Because data sets B_1 and B_2 are on same document it appears a change affecting A would also affect C. This error is removed by breaking B into documents B_1 and B_2.

The third problem concerns timing. A document may include several pieces of closely related information. One of these pieces may be produced by a task performed early in the project and needed by another task early in the project. However, other pieces of information on the same document may not be available until much later in the project. If the publication of the document is held up for the availability of these later pieces of information, then the first item of information will not be documented when it is needed. This is represented in Figure 7.7. This problem disappears if one uses a computerized data base, because it is no longer necessary to have the publication of a document synchronize the availability of data.

In most organizations the shortcomings of the formal documentation system are overcome by an informal, undocumented flow of information.

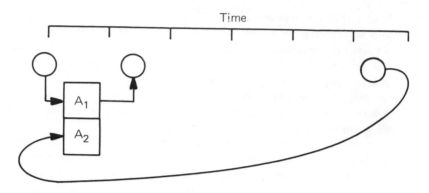

Figure 7.7
Scheduling Error Due to Document Definition. Data set produced late in project holds up document needed early in project.

When information is needed after it becomes available informally but before it can be documented, then that information will flow by word of mouth or on scraps of paper. The process of designing many systems would grind to a halt if it were not for such informal information transfers. However, these informal transfers can generate serious problems of their own. Different people may be working with different values for the same data. When assumptions are not documented, they may not get verified later when the required basis for that data is available. It becomes difficult or impossible to control what information people are using and to ensure that what data they do use is properly verified. One cannot then trace the basis for the final design. This creates a problem in justifying the design to the customer or a regulatory agency, or using the design as a basis for another similar one. These difficulties can be avoided to the extent that it is possible to design the formal information flow to decrease the burden on the informal information flow.

Another phenomenon which is not uncommon is the existence of a number of different documents containing the same data, and often not the same values for that data. This can be the source of much confusion. Frequently, a document will include values for data that were assumed by the author in the absence of any better values for that data. Authors of other documents may use these same assumptions, or sometimes develop their own. The design review process may verify some of these estimates, not verify others, and be completely unaware of the use of yet other estimates. It sometimes means that the whole design process has to be run through again to check all assumptions.

The design structure matrix shows the design process, where assumptions are made, and where they are used. It should then be possible to set up documents showing exactly what values are used for the assumptions, and when and by whom they are to be reviewed and verified. If all the design work follows from well-documented assumptions, then once the reviews at the end of the blocks establish the validity of those assumptions, the design tasks of that block are finished. If certain assumptions are not verified, then one can use the techniques discussed earlier to trace exactly which other tasks are affected and then make or revise plans and schedules accordingly.

7.8 The Collection of Precedence Data

Given a complete set of well-defined tasks and individual responsibilities for those tasks, it is comparatively straightforward to collect the precedence constraints and do the analysis required to set up a design structure matrix. Each engineer responsible for a task should be able to tell you which of the set of other tasks are needed to do the work. If the engineer cannot give you this information, there is good reason to doubt whether he can perform the task at all.

Note, however, that while an individual responsible for a task may be able to tell you what all the predecessors are, he probably could not tell you what all the successors are. He may know some of the uses for the results of the task since he will usually have an idea of what the basic motivation for doing the task is. However, he usually will not know all of the uses to which the results are put. But it is the successors that he needs to know in order to distribute the outputs and the consequences of any change in the task to the correct people. Fortunately, it is quite sufficient to ask him what the predecessors are, for knowing these for each task in the system it is then only a simple data processing procedure to invert this list to obtain the successors for each task.

Before the precedence constraint data can be gathered, it is necessary to have defined a complete set of tasks. This definition is the most difficult part in the process of setting up and analyzing a design structure matrix.

One problem which arises immediately is how finely the tasks should be divided. Figure 7.8 shows a flowchart of the process associated with doing a piece of engineering work leading to the production of a drawing or other document. Should each of these individual operations be treated as a separate task, or should the set of operations be wrapped up as one task? Should a task represent the engineering required to define all the data on a particular document, or should there be separate tasks for the engineering required to define each subset of that data, or even each data item?

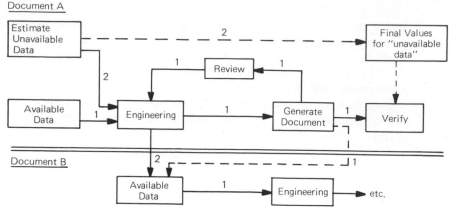

1 = Standard transmittal
2 = Nonstandard transmittal

This detail may be summarized as:

Figure 7.8
Detail Flowchart for Production of an Engineering Document

The finer the detail in defining the tasks the greater the effort required to collect and analyze the data, but the greater also are the benefits to be derived from the analysis. In some modest-sized engineering operations, if a task were used to represent the determination of each data value there would be tens to hundreds of thousands of tasks. But as the next section shows, if we break up some of the tasks so they represent individual pieces of data, we can make useful decisions about the design of the documents, e.g., which combination of data elements should occur on the same document. This trade-off of detail versus added work to collect and analyze the data must be evaluated for each individual application.

One approach to working with large systems is to first define a small number of large tasks to make a preliminary analysis. Then based upon this analysis only tasks which warrant further analysis need be broken into more detailed tasks. Let us first consider this top-down approach.

We begin with the assumption that there exists some means of identifying a complete set of documents. By a complete set of documents we mean that once all of these documents are generated we can say that the design is complete. If we do not already have a complete set of documents identified, then as a first step we must identify such a set. The next section will be concerned with whether these documents are designed correctly to contain the right collection of information on each document. But at this stage we will be satisfied to use whatever set of documents we have or can define. We must also assign a responsible individual to each document.

Then we can distribute to each individual responsible for a document a form such as in Figure 7.9. This form asks the person to list the other documents and sources from which he obtains the data required as a prerequisite to producing this document. It may be necessary to supply or make available to the engineer a list of all the documents in the system he is likely to use with their proper document designations so that all individuals refer to the same document in the same way. Using this information for all the documents, one can construct a precedence matrix showing document-to-document relations.

The same form also asks for the definition of the data sets which appear on the document. A data set is a set of data which is similarly defined, similarly derived, and similarly used. For example, a set of values giving the power density at each point in the core of a nuclear power plant under certain conditions could represent a data set. Two different individuals might define the data sets on a particular document somewhat differently. We propose that one person, the individual responsible for the document, define the data sets for that document. Everyone else should use these definitions consistently unless they can show a good reason why these definitions should be changed. Such a reason might be that a data set should be broken down because various subsets are used in different ways or for different tasks. The information collected in this way is used to develop forms for collecting the more detailed data set-to-data set relations.

If it is decided that the relations between certain documents should be broken down to the level of the data sets on the document for a more detailed analysis, one can then collect the information on a form such as shown in Figure 7.10. Using the information from the earlier form, a tailor-made form is produced by the computer showing the data sets which occur on each document and the data sets which occur on each of the predecessor documents. (The data sets for the predecessor documents are those specified on the earlier form by the individual responsible for those predecessor documents.) The engineer responsible for the subject document can then check off which data sets from the predecessor documents are required for each data set appearing on the subject document. The use of these forms based upon the information collected on the earlier form helps to collect the data so the tasks, either the documents or data sets, are defined consistently. Using this information a precedence matrix can be formulated and analyzed based upon data set-to-data set relations. The next section discusses some of the conclusions about the structure of the documents which can be derived from these data set-to-data set relations.

DOCUMENT NUMBER   _E 5013_
DOCUMENT TITLE   _Heat Transfer Parameters_
RESPONSIBLE ENGINEER   _Robert Smith_
RESPONSIBLE COMPONENT   _185 Thermo Analysis Section_
ESTIMATED MAN-HOURS 1st __70__ , 2nd __50__ , and nth __45__

DATA SETS ON THIS DOCUMENT   DATA SET #
                                              (.1, .2, etc.)

SFC H/T COEF	.01
MAX HEAT FLX	.02
AVE HEAT FLX	.03
H/T AREA/BDL	.04
POW DENSITY	.05
MCHFR	.06
AVE SPEC POW	.07
GAP H/T COEF	.08

PREDECESSOR DOCUMENTS REQUIRED

Document #	Document Title
A	BASIC PLANT PARAMETERS
C	FUEL-PHYSICS
D	HYDRAULIC - FLUID FLOW
F	CORE GEOM & PHYSICS
H	CHEMICAL SYSTEMS

Figure 7.9
Data sets and predecessor documents—filled-out input form

```
PAGE  1
EXTERNAL IDENT = E5013
DESCRIPTION    = HEAT TRANSFER PARAMETERS
RESP. ENGINEER = ROBERT SMITH
RESP. COMP.    = 185 THERMO ANALYSIS SECTION
```

DATA SETS ON THIS DOCUMENT	MAN-HR	NUMBER	01	02	03	04	05	06	07	08
SFC H/T COEF	5	F.01	*						X	
MAX HEAT FLX	10	E.02			*	X			X	
AVG HEAT FLX	10	E.03		X		*		X		
H/T AREA/BDL	10	E.04					*	X		
POW DENSITY	5	E.05						*		
MCHFR	15	E.06							*	
AVG SPEC POW	10	E.07		X					*	
GAP H/T COEF	5	F.08							8	*

Figure 7.10a

Computer-Generated Data Set-to-Data Set Input Form Filled Out

```
PAGE  2
EXTERNAL IDENT = E5013
DESCRIPTION    = HEAT TRANSFER PARAMETERS
RESP. ENGINEER = ROBERT SMITH
RESP. COMP.    = 185 THERMO ANALYSIS SECTION
```

DATA SETS ON THIS DOCUMENT	NUMBER	01	02	03	04	05	06	07	08
SFC H/T COEF	E.01	*							
MAX HEAT FLX	E.02		*						
AVG HEAT FLX	E.03			*					
H/T AREA/BDL	F.04				*				
POW DENSITY	E.05					*			
MCHFR	F.06						*		
AVG SPEC POW	E.07							*	
GAP H/T COEF	E.08								*

DATA SETS ON PREDECESSOR DOCUMENTS

A BASIC PLANT PARAMETERS

	NUMBER	01	02	03	04	05	06	07	08
MWE	A.01								
MWT	A.02								
PRESSURE-SAT	A.03	X					8	9	
FEEDWAT-TEMP	A.04						8		
STEAM FLOW	A.05								
SEC.SYST&TUR	A.06								

Figure 7.10b

```
    PAGE   3
EXTERNAL IDENT  = E5013
DESCRIPTION     = HEAT TRANSFER PARAMETERS
RESP. ENGINEER  = ROBERT SMITH
RESP. COMP.     = 185 THERMO ANALYSIS SECTION
```

DATA SETS ON THIS DOCUMENT	NUMBER	01	02	03	04	05	06	07	08
SFC H/T COEF	E.01	*							
MAX HEAT FLX	E.02		*						
AVG HEAT FLX	E.03			*					
H/T AREA/BDL	E.04				*				
POW DENSITY	E.05					*			
MCHFR	E.06						*		
AVG SPEC POW	E.07							*	
GAP H/T COEF	E.08								*

DATA SETS ON PREDECESSOR DOCUMENTS

C FUEL-PHYSICS

	NUMBER	01	02	03	04	05	06	07	08
MOD/FUEL VOL	C.01								
RODS/BUNDLE	C.02				X				
CLAD THK/DIA	C.03								
CONT. PITCH	C.04					X			
CHAN. THICK	C.05								
WIDE GAP	C.06								
NARROW GAP	C.07								
UO2 DENSITY	C.08		X					X	
EXPOSURE	C.09								
FUEL PROCESS	C.10								
UO2 DIAM	C.11		X						
CLAD THICK	C.12								
PELLET-CLAD	C.13								
FUEL ROD O.D	C.14		X		X			X	
CENTTEMPLIM	C.15							X	
MOD HEAT FR.	C.16		8			8			
CONTROL SPAN	C.17								

Figure 7.10c

```
     PAGE   4
  EXTERNAL IDENT  = E5013
  DESCRIPTION     = HEAT TRANSFER PARAMETERS
  RESP, FNGINEER  = ROBERT SMITH
  RESP, COMP,     = 185 THERMO ANALYSIS SECTION
```

DATA SETS ON THIS DOCUMENT	NUMBER	01	02	03	04	05	06	07	08
SFC H/T COEF	E.01	*	.	.	.	.	.	.	.
MAX HEAT FLX	E.02	.	*	.	.	.	.	.	.
AVG HEAT FLX	E.03	.	.	*	.	.	.	.	.
H/T AREA/BDL	E.04	.	.	.	*	.	.	.	.
POW DENSITY	E.05	.	.	.	.	*	.	.	.
MCHFR	E.06	.	.	.	.	.	*	.	.
AVG SPEC POW	E.07	.	.	.	.	.	.	*	.
GAP H/T COEF	E.08	.	.	.	.	.	.	.	*

DATA SETS ON PREDECESSOR DOCUMENTS

D HYDRAULIC-FLUID FLOW

		01	02	03	04	05	06	07	08
BYPASS FRACT	D.01	.	.	.	.	.	.	.	.
SEP,CARY/UND	D.02	.	.	.	.	.	.	.	.
HYDR, DIAM,	D.03	.	.	.	.	.	.	.	.
ORIFICE PATT	D.04	.	.	.	.	.	8	.	.
FLOW/BUNDLE	D.05	.	.	.	.	.	8	.	.
TOTAL FLOW	D.06	.	.	.	.	.	.	.	.
EXIT QUALITY	D.07	.	.	.	.	.	X	.	.
AVG VOIDS	D.08	.	.	.	.	.	.	.	.
STEAM SEPFLO	D.09	.	.	.	.	.	.	.	.
JET PUMP EFF	D.10	.	.	.	.	.	.	.	.
CORE PRESDRP	D.11	.	.	.	.	.	.	.	.
ORIF PRESDRP	D.12	.	.	.	.	.	.	.	.

Figure 7.10d

```
        PAGE   5
    EXTERNAL IDENT  = E5013
    DESCRIPTION     = HEAT TRANSFER PARAMETERS
    RESP. ENGINEER  = ROBERT SMITH
    RESP. COMP.     = 185 THERMO ANALYSIS SECTION
```

DATA SETS ON THIS DOCUMENT	NUMBER	01	02	03	04	05	06	07	08
SFC H/T COEF	E.01	*	·	·	·	·	·	·	·
MAX HEAT FLX	E.02	·	*	·	·	·	·	·	·
AVG HEAT FLX	E.03	·	·	*	·	·	·	·	·
H/T AREA/BDL	E.04	·	·	·	*	·	·	·	·
POW DENSITY	E.05	·	·	·	·	*	·	·	·
MCHFR	E.06	·	·	·	·	·	*	·	·
AVG SPEC POW	E.07	·	·	·	·	·	·	*	·
GAP H/T COEF	E.08	·	·	·	·	·	·	·	*

DATA SETS ON PREDECESSOR DOCUMENTS

F CORE GEOM & PHYSICS

	NUMBER	01	02	03	04	05	06	07	08
CORE VOL.	F.01	·	·	·	·	·	·	·	·
CORE RADIUS	F.02	·	·	·	·	·	·	·	·
CORE LENGTH	F.03	·	·	·	X	X	·	·	·
LOCAL P/A	F.04	·	·	·	·	·	·	·	·
AXIAL P/A	F.05	·	·	·	·	·	X	·	·
RADIAL P/A	F.06	·	·	·	·	·	·	·	·
TOTAL P/A	F.07	·	·	X	·	·	·	X	·
VOID COEFF	F.08	·	·	·	·	·	·	·	·

Figure 7.10e

```
        PAGE   6
    EXTERNAL IDENT  = E5013
    DESCRIPTION     = HEAT TRANSFER PARAMETERS
    RESP. ENGINEER  = ROBERT SMITH
    RESP. COMP.     = 185 THERMO ANALYSIS SECTION
```

DATA SETS ON THIS DOCUMENT	NUMBER	01	02	03	04	05	06	07	08
SFC H/T COEF	E.01	*	·	·	·	·	·	·	·
MAX HEAT FLX	E.02	·	*	·	·	·	·	·	·
AVG HEAT FLX	E.03	·	·	*	·	·	·	·	·
H/T AREA/BDL	E.04	·	·	·	*	·	·	·	·
POW DENSITY	E.05	·	·	·	·	*	·	·	·
MCHFR	E.06	·	·	·	·	·	*	·	·
AVG SPEC POW	E.07	·	·	·	·	·	·	*	·
GAP H/T COEF	E.08	·	·	·	·	·	·	·	*

DATA SETS ON PREDECESSOR DOCUMENTS

H CHEMICAL SYSTEMS

	NUMBER	01	02	03	04	05	06	07	08
RESIDENCE TM	H.01	·	·	·	·	·	·	·	·
CRUD RATE	H.02	·	·	·	·	·	·	9	·

Figure 7.10f

Another approach to building precedence matrices for large systems is through merging of the precedence matrices collected for the subsystems. Figure 7.11 illustrates this bottom-up approach. The subsystems may be well-defined subsystems derived from the structure of the product (i.e., a parts list explosion defining the hardware for the system) or they may be defined as those tasks which presently are the responsibility of a particular component of the organization. As seen in Figure 7.11, the precedence matrix for each subsystem may involve many tasks in other subsystems. The same task may be repeated in many matrices.

Let us assume that we have obtained precedence matrices for each of a number of organizational components. Each matrix will contain the tasks assigned to that component plus those tasks assigned to other components which are the direct predecessors of tasks in this component. The tasks from the other components will move to the beginning of the matrix when it is partitioned and torn.

Now it is possible to merge all these matrices into one large composite matrix. However, we may find that this is too many tasks to visualize and comprehend. Thus it is possible to collapse—that is, combine—certain sets of tasks such that each collapsed set of tasks in a component matrix becomes a single task in the composite matrix.

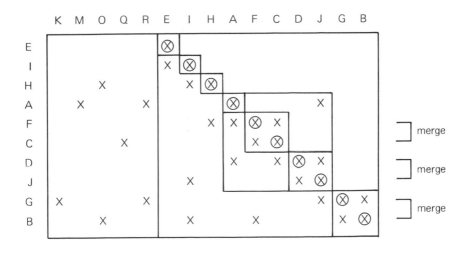

(a)

Figure 7.11
Merging of Design Structure Matrices. (a) Design structure matrix for Unit 183. (b) Design structure matrix for Unit 297. (c) Merged design structure matrix.

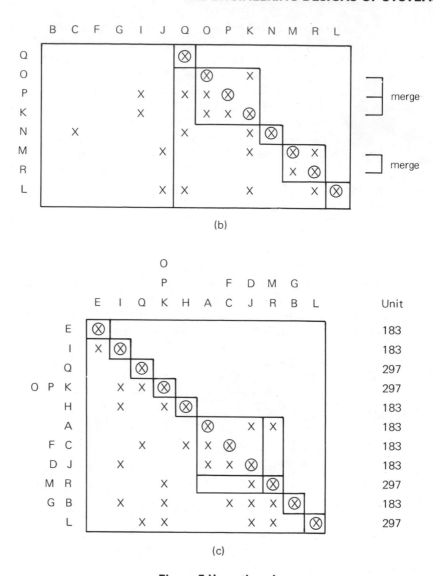

(b)

(c)

Figure 7.11 continued

Now the question arises of how to choose the sets of tasks to collapse. An engineer analyzing each component matrix might collapse sets of tasks which are closely related and can be defined easily as a group.

It might be quite sensible to collapse those sets of tasks in a component matrix which occur together in a common block in the partition after all marks of higher than a given level, possibly zero, have been removed. This approach then parallels Procedure 3.2 for partitioning in which the set of tasks in a circuit is collapsed before searching for another circuit.

7.9 The Strategy for Tearing

Before discussing the strategy of tearing, let us hasten to point out that it is not necessary that every engineer know how to do this tearing analysis. This can be done by one analyst or engineer who has some understanding of the engineering work and a good understanding of the process of tearing. He will take the precedence data and make a tearing analysis to obtain a proposed design structure matrix, which he will review with the engineers responsible for the work. As they criticize the engineering plan implied by this design structure matrix, he will show them other alternatives based upon his knowledge of the tearing process. In this way a final design structure matrix is arrived at. Each engineer does not have to understand how the design matrix was arrived at, provided he understands how to interpret and use it.

With this in mind we shall proceed to discuss in this and the following sections some of the lore expected of this analyst.

Tears represent where the results of tasks not yet performed must be assumed to start the iteration process.

We wish to order the tasks in each block so that the iterative process produces an acceptable design with the least amount of iteration. In section 5.3 we saw that for an iteration around a single circuit in a linear system of equations, we could state a rather simple criterion for the rate of convergence of the iteration: On each iteration the error is multiplied by the product, with appropriate sign, of the derivative of each row variable in the circuit with respect to its predecessor. If the product was less than 1 in absolute value, the process converged. If the product was greater than 1 in absolute value, the process diverged. By reversing the direction of the iteration, which can be done by changing the tearing, the reciprocal of this product is produced. Thus, if the process diverged in one direction, it converged in the opposite direction.

It is not usually possible to state a general engineering design process in such pure and simple terms. The process is not usually represented by simple linear equations. However, this suggests that it is preferable to tear where the task is insensitive to the uncertainty in the a priori knowledge of the predecessor. This means that we will be more inclined to accept an assumption (tear) to the extent that we have a favorable combination of the following conditions: (1) the error in the assumed value of the predecessor is likely to be small, probably because of prior experience with similar systems, and/or (2) the task itself is not sensitive to errors in the predecessor. We state this condition rather imprecisely because the conditions of its application are usually not very accurate.

Subject to these conditions, which pertain to interpreting the semantics represented by the precedence relations, we can make some general statements about how we might choose a set of tears.

Section 4.2 discussed four possible criteria for choosing tears. We can now consider these criteria from the point of view of their application to the engineering design of systems.

Our experience indicates that the most pertinent criterion is: "Confine the most tears to the smallest square block on the diagonal." This leads to a design structure matrix in which smaller blocks are iterated within larger blocks. The communication which occurs during the iteration of the smaller blocks can be confined to a smaller number of persons. Often, some care in these less expensive inner iterations can reduce the number of more expensive outer iterations.

Outside these small inner blocks we may attempt to use the criterion, "The tears should appear in the least number of different rows." These rows will then represent the tasks or documents where the assumptions are made. Thus, these tasks or documents will be the responsibility of the more experienced persons. The assumptions represented by these tears should be documented clearly so that they can be reviewed at the end of the block.

We also look to see if there is a good tearing which has tears in a small number of different columns, meaning the number of different tasks whose results must be assumed is small. Or we may seek just a small number of tears, which implies that we reduce the number of places that we introduce errors due to using assumptions.

There is no simple, rigorous criterion that we use for a best tearing. Each of the four criteria proposed in section 4.2 has some merit. We often look at all of them; if we find a tearing which meets one of these criteria particularly well, it usually behooves us to look carefully at the engineering interpretation of that particular tearing. Many times we generate several tearings for interpretation before settling on one which looks good.

Often we will tear those precedence relations which apply to optimality in preference to tearing the relations pertaining to feasibility. By feasibility we refer to those relationships which affect whether the system will work at all or work acceptably. By optimality we refer to those relationships which pertain to how effectively or inexpensively the system will work. We generally expect than if we make a change to a system and demand only that the consequent changes guarantee feasibility, then we are less likely to be faced with circuits than if we were to demand that the new design be optimized for the changed conditions.

Sometimes we purposely introduce variables, called *conformance variables,* which are intended to compare the computed performance of a design with the performance required by the specifications, and feed this comparison back to those control variables which can affect this comparison. The design of an electric car discussed in chapter 2 contains such a variable, variable 11, Speed and Acceleration Conformance, which compares the Cruising Speed Specification and Acceleration Specification, on the one hand against Speed and Acceleration Performance versus Power on the other hand, to feed back the difference to the control variables Size-Aerodynamics, Total Weight, and Structural and Suspension Design. These control variables might then be changed to affect the comparison. One knows from the formulation of the problem that the use of such a conformance variable is expected to be a feedback. Thus, the marks in the column associated with such a variable will ordinarily be given a high-level number, e.g., 9, in the tearing process.

7.10 Data Processing Techniques for Collecting and Analyzing the Data

The handling of the data collection and the administering of the system recounted in this chapter could be facilitated by the use of a system of computer programs. (The reader is referred to section 2.9 to see how these programs relate to distributed processing.)

Let us describe the basic programs in the system. Two of these programs have already been considered, TERABL (see the Appendix) and ASPECT.

A. EXTERNAL TO INTERNAL NUMBER ASSIGNMENT

Each task is given an identifying external identification. For internal use within the system, we must assign to the tasks the integers from 1 to n where there are n distinct tasks.

We maintain source data by their external identifications. These external identifications need not be changed if other tasks are added or deleted.

B. TERABL

The TERABL program, discussed in the Appendix, accepts the output from program A. It is used to partition and tear the system, producing a new ordering. The order table it generates can be used in Program C.

C. RENUMBER AND SORT

This program renumbers the internal task numbers according to the order table produced by the TERABL program. The order table shows the correspondence between the old and the new internal numbers. The records are then sorted in ascending order of these new numbers. This program can also be used to reorder tasks for printing out the matrix to show the consequences of various orderings.

   The task input and output of program C corresponds in format to the output of A. The renumbering table format corresponds to the output of TERABL.

D. GENERATE DATA SET–TO–DATA SET INPUT FORM

This program generates an input form to be filled out by an engineer as shown in Figure 7.9.

E. INVERSION OF LISTS

This program accepts a precedence list and produces a successor list. Both the input and output have the format of the output program A. The output of this program is used to set up the files used by ASPECT. In addition, an output is produced showing for each task the internal number, identification and man-hour information from the output of program A, a list of the predecessors and a list of the successors and their identifications. The predecessors or successors involved in circuits are flagged with an asterisk. This output, called an IMPACT list, is used to replace the matrix coming from the TERABL program for those managers who do not care to learn how to interpret the matrices.

F. ASPECT

Files are set up on disc storage showing predecessors and successors, descriptions, responsible individuals and organizations, and man-hour estimates for the tasks. The ASPECT program can then be used to determine from a time-shared terminal all the tasks affected by specified changes and to retrieve the descriptions, responsible individuals and organizations, and man-hour estimates for the tasks affected. A task list of the subset of tasks affected by the changes is generated for the use of programs G and H.

G. INTERFACE TO TERABL

This program accepts as input from ASPECT the list of tasks affected by the specified changes. It then generates input to TERABL, assigning a new set of internal numbers for the subset of tasks. TERABL can then be used for an analysis of the structure of the design process for just this subset.

H. INTERFACE TO IMPERT

This program accepts the same input as required for the interface to TERABL, plus the task ordering obtained from running TERABL after ASPECT, plus a list showing for each block the decision reached for how many times it is to be iterated. A block is described by giving the internal numbers of the first and last tasks in the block. The output, which becomes the input to the IMPERT critical path scheduling program, gives the precedence relations with the circuits unwrapped.

I. IMPERT

IMPERT is a critical path scheduling program whose input is based upon listing the predecessors for each task rather than the PERT event-node format.

J. MAN-LOAD

The man-loading program acepts the output of IMPERT and a list of the available time for each man involved.*

Figure 7.12 represents this system of data processing programs. Part (a) shows the information which appears on the various input forms; part (b) illustrates the layouts of the data files. It should be noted that a common input data file layout (D1) is used by most of the programs in this system. This allows these programs to be used in very versatile ways in different systems. The descriptors show how the information in that basic data file layout changes at various places in the system. Part (c) exhibits the table of reports. Parts (d) through (g) show how these programs can be used to expand document precedence relations into data set precedence relations. Parts (h) and (i) demonstrate how the same set of programs can be used to collapse tasks and merge matrices to obtain summary matrices.

*See Steward:67c for a detailed discussion.

Figure 7.12a
Table of Input Forms

I1: Input Form 1
  Document Number of Subject Document
  Document Title
  Responsible Engineer
  Responsible Component
  Estimated Man-Hours: 1st, 2nd, and nth iteration
  For each Data Set of Subject Document:
    Description of Data Set
    Data Set Number (may be a decimal number, e.g., .1, .2, etc., to be appended to document number)
  For each Predecessor Document:
    Document Number
    Document Title

I2: Input Form 2
  Document Number of Subject Document @
  Document Title @
  Responsible Engineer @
  Responsible Component @
  For each Data Set Appearing on Subject Document:
    Man-Hours to Produce this Data Set
    (Respondent marks predecessors from data sets within subject document)
  For each Predecessor Document:
    Data Sets on Subject Document
    (Respondent marks predecessor data sets from within this predecessor document)
  @ implies information preprinted by computer

I3: Input Form 3
  For each Block:
    Identification of the Block by giving internal numbers of first and last tasks of the block
    Estimated Number of Iterations of the Block

I4: Input Form 4
  Used to preset table in Program A assigning internal numbers to external identifications so as to collapse tasks with different external identifications into a task with a common internal number.
  For each new task formed from the collapsing of tasks:
    List of External Identifications to be collapsed into this task
    Description of Collapsed Task
    Responsible Engineer
    Responsible Component
    Estimated Man-Hours: 1st, 2nd and nth iterations

Figure 7.12b
Table of Data Files

D1: Data 1
    External Identification of Document or Data Set
    Internal Identification (may be blank)
    Document or Data Set Description
    Responsible Engineer
    Responsible Component
    Estimated Man-Hours: 1st, 2nd and nth iteration
    For each Predecessor or Successor Document or Data Set:
        External Identification
        Internal Identification (may be blank)
    Descriptors:
        n (without internal identification) or i (with internal identification)
        p (predecessors) or s (successors)
        c (course—documents) or f (fine—data sets)
        l (little-subset)
        o (ordered by partitioning and tearing)
D2: Data 2
    External Identification of Document
    For each Data Set appearing on Document:
        External Number of Data Set
        Description of Data Set
D3: Data 3
    List of integers defining the reordering of internal identification numbers

Figure 7.12c
Table of Reports

R1: Report 1
    Printed output from TERABL including Design Structure Matrix
R2: Report 2
    Printed IMPACT list:
        For each Document or Data Set:
            List of Predecessors
            List of Successors
R3: Report 3
    Printed output from IMPERT
        For each Activity:
            Activity Number   Earliest Finish
            Activity Description   Latest Beginning
            Duration   Latest Finish
            Iteration Number   Slack
            Earliest Beginning
        Printed Bar Chart
R4: Report 4
    Man-Load Reports
        For each Engineer:
            For each Week:
                For each Activity to be worked on that week:
                    External Identification
                    Estimated Man-Hours to be worked

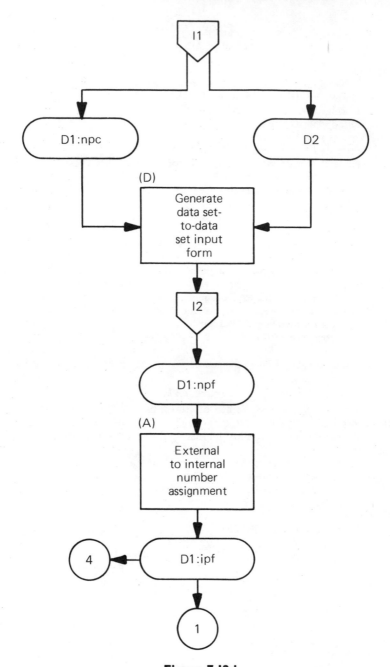

Figure 7.12d
Flowchart Representing Expansion of Documents into Data Sets and Genera-
tion of Schedules, etc.

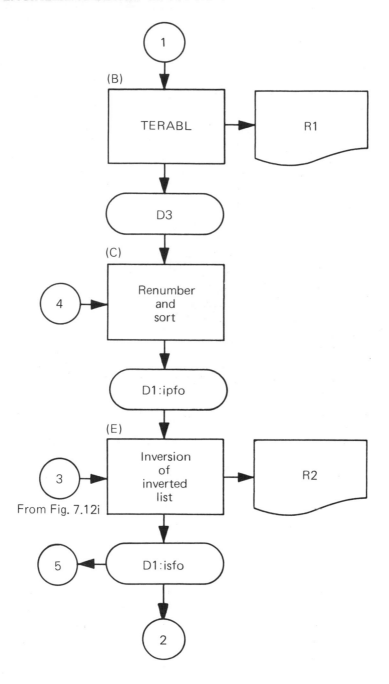

Figure 7.12e

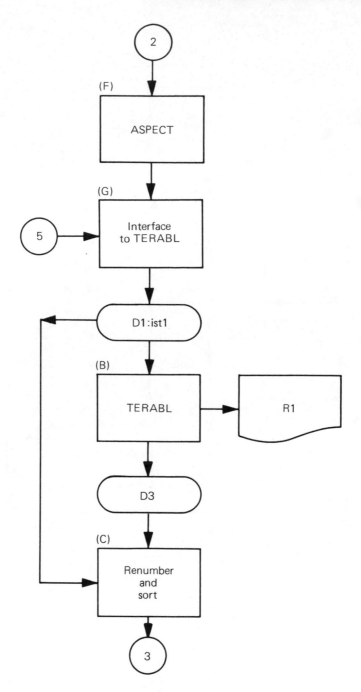

Figure 7.12f

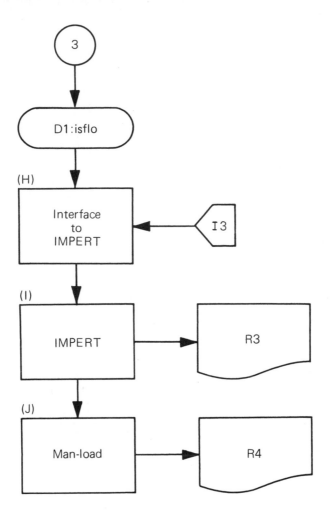

Figure 7.12g

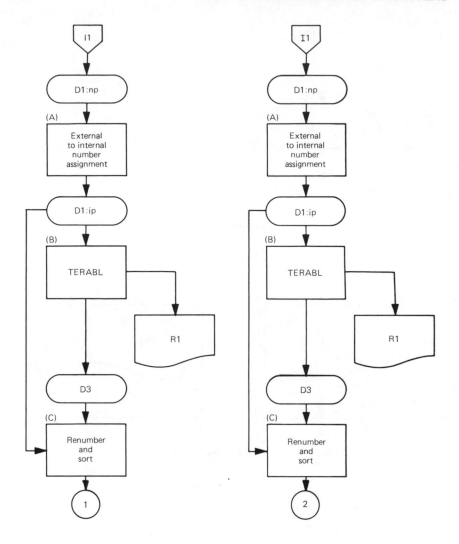

Figure 7.12h
Flowchart Representing Collapse of Tasks and Merging of Matrices

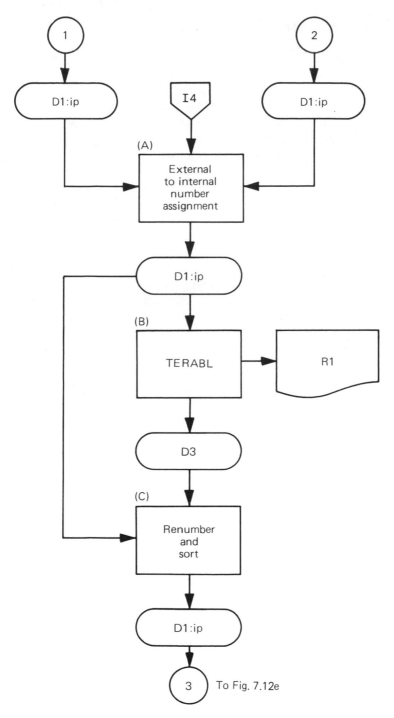

Figure 7.12i

7.11 An Example

Since this section refers to the output of the TERABL program, a brief preliminary reading of the Appendix describing the program and its outputs might first be in order. We shall now apply the techniques discussed in this chapter to a fictionalized example. See also Chapter 4.

Figure 7.9 showed a form filled out by a hypothetical Robert Smith, an engineer responsible for the heat transfer document. On this form he has shown (1) what other documents he needs data from before he can finish his heat transfer document, (2) the data sets which appear on his document, and (3) the estimated time he requires to produce his document. Figure 7.13 shows the precedence matrix obtained by putting together the precedence data for all of the documents. Figure 7.14 shows the design structure matrix obtained by partitioning and tearing to obtain a viable design procedure.

Documents

Documents	A	B	C	D	E	F	G	H	I
A	⊗	9							
B	X	⊗		X					
C	X		⊗	X	X			X	X
D	X		8	⊗	8	8	8	X	
E	X		8	X	⊗	8		X	
F	X		X	X	X	⊗			
G	X		X	X	X	X	⊗	X	
H								⊗	
I									⊗

Figure 7.13
Precedence Matrix Based on Documents

Now we wish to go to a greater level of detail by breaking these documents into their data sets and obtain the precedence relations between the data sets. Using the information from forms such as Figure 7.9 collected for all documents, the computer produces a new form for Mr. Smith (see Figure 7.10). This form shows the data sets he had indicated appear on his document and the list of data sets from each predecessor document. Using this form Mr. Smith has shown for each data set appearing on his document what data sets from his and other documents

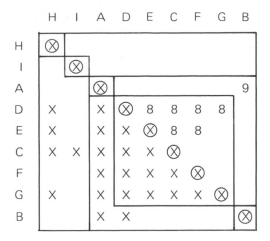

Figure 7.14
Design Structure Matrix Based on Documents

are required as predecessors. Note that he has used 8's and 9's to indicate those predecessors which he believes are less sensitive to errors in the a priori knowledge of the predecessors, i.e., either there is a good basis for estimating the predecessor, or a large error in the predecessor can be tolerated, or both. Figure 7.15a shows the precedence matrix for the data set-to-data set relations obtained by putting together the information obtained from these forms for all of the documents. Figure 7.15b shows the size of the blocks in the partition.

Part (c) Figure 7.15 shows this precedence matrix partitioned by levels as per Procedure 3.4. Initially, there is a large block of size 23. When the 9's are removed, this block partitions further, leaving smaller blocks of sizes 2, 6, and 11 within the larger blcok. Removing the 8's causes the block of size 11 to partition yet further showing a block within it of size 9. Note that a number of 8's and 9's fall below the diagonal indicating that they were not necessary for breaking all the circuits. The outlines of the blocks have been drawn by hand on the computer printout.

Now we wish to consider how we are going to order the tasks within the blocks which remain after the 8's and 9's are removed. Let us first consider the block of size 2. It is so simple that a shunt diagram is not required and is not printed; this block involves just Fuel Rod O.D. and the Pellet-Clad volume ratio. We must initially assume one in order to determine the other. It would probably make more engineering sense to assume the Pellet-Clad volume ratio first as we have a good estimate of this from earlier designs and any likely error would not significantly affect the

			#	1	2	3	4	5	6	7	8	9	10	11	12	13	14	15	16	17	18	19	20
1	A	MWF	1	X																			
1	A	MWT	2	X	X					X													
1	A	PRESSURE-SAT	3			X																	
1	A	FEEDWAT-TEMP	4			X																	
1	A	STEAM FLOW	5		X	X	X	X															
1	A	SEC,SYST&TUR	6	X		X	X		X	8													
2	B	PUMPING POW	7			X		X		X													
2	B	THERM EFFIC.	8			X	X		8	X	X												
3	C	MOD/FUEL VOL	9									X											
3	C	RODS/BUNDLE	10										X										
3	C	CLAD THK/DIA	11											X									
3	C	CONT. PITCH	12												X								
3	C	CHAN. THICK	13												X	X	8	8					
3	C	WIDE GAP	14													X							
3	C	NARROW GAP	15															X					
3	C	UO2 DENSITY	16																	X	X		
3	C	EXPOSURE	17																	X			
3	C	FUEL PROCESS	18																		X		
3	C	UO2 DIAM	19																			X	
3	C	CLAD THICK	20											X									X
3	C	PELLET-CLAD	21		8																	X	
3	C	FUEL ROD O.D	22									X	X	8	X	9							
3	C	CENTTEMPLIM	23																				
3	C	MOD HEAT FR.	24									X		X		X	X						
3	C	CONTROL SPAN	25											X									
2	D	BYPASS FRACT	26																				
2	D	SEP,CARY/UND	27																				
2	D	HYDR. DIAM.	28									X		8		9	9						
2	D	ORIFICE PATT	29																				
2	D	FLOW/BUNDLE	30			X								X									
2	D	TOTAL FLOW	31			X																	
2	D	EXIT QUALITY	32																				
2	D	AVG VOIDS	33		X	X																	
2	D	STEAM SEPFLO	34			X																	
2	D	JET PUMP EFF	35																				
2	D	CORE PRESDRP	36									X		X	8	X	X						
2	D	ORIF PRESDRP	37																				
2	E	SFC H/T COEF	38			X																	
2	E	MAX HEAT FLX	39																	X		X	
2	E	AVG HEAT FLX	40																				
2	E	H/T AREA/BDL	41									X											
2	E	POW DENSITY	42											X									
2	E	MCHFR	43			8	8																
2	E	AVG SPEC POW	44		9															X			
2	E	GAP H/T COEF	45																				
3	F	CORE VOL.	46	X																			
3	F	CORE RADIUS	47																				
3	F	CORE LENGTH	48																				
3	F	LOCAL P/A	49																				
3	F	AXIAL P/A	50																				
3	F	RADIAL P/A	51																				
3	F	TOTAL P/A	52																				
3	F	VOID COEFF	53		X							X											
3	G	VOID TRANTIM	54			8	X																
3	G	FUELTIMCONST	55																	X		X	X
3	G	STABIL.CRIT.	56																				
4	H	RESIDENCE TM	57																				
4	H	CRUD RATE	58																				
1	I	CAP. FACTOR	59																				
1	I	REFUEL PLAN	60																				

Figure 7.15a
TERABL Output—Thermo-Hydraulics-Physics

THERMO-HYDRAULICS-PHYSICS
ORDER 60
HIGHEST LEVEL MARK TO BE CONSIDERED   9
LARGEST BLOCK FOR WHICH SHUNT DIAGRAM IS NOT PRINTED   2
LARGEST INDEX SHUNT TO BE CONSIDERED   60

LEVEL   9

```
  1  1  1  1  1  1  1  1  1  1  1  1  1  1  1  1  1  1  1  1
  1  1  1  1  1  1  1  1  1 23  1  1  1  1  1  1  1  1  1
```

Figure 7.15b

LEVEL 8
1 1 1 1 1 1 1 1 1 1 1 1 1 1 1 1 1 1 1
1 1 1 1 1 1 1 1 1 2 1 1 6 1 1 11 1 1 1 1
1 1 1 1

LEVEL 0
1 1 1 1 1 1 1 1 1 1 1 1 1 1 1 1 1 1 1 1
1 1 1 1 1 1 1 1 1 2 1 1 6 1 1 9 1 1 1 1
1 1 1 1 1 1

```
THERMO-HYDRAULICS-PHYSICS                                    AFTER PARTITIONING
     PAGE  1/ 1
                                    0 0 0 0 0 0 0 0 0 0 0 0 0 0 0 0 0 0 0 0
                                    0 0 0 0 1 1 1 1 1 2 2 2 2 2 3 3 4 4 4
                                    1 3 4 9 0 1 2 4 5 8 3 4 5 6 7 2 5 5 8 9

  1 A   MWE            1   X . . . . . . . . . . . . . . . . . . .
  1 A   PRESSURE-SAT   3   . X . . . . . . . . . . . . . . . . . .
  1 A   FEEDWAT-TEMP   4   . . X . . . . . . . . . . . . . . . . .
  3 C   MOD/FUEL VOL   9   . . . X . . . . . . . . . . . . . . . .
  3 C   RODS/BUNDLE   10   . . . . X . . . . . . . . . . . . . . .
  3 C   CLAD THK/DIA  11   . . . . . X . . . . . . . . . . . . . .
  3 C   CUNT. PITCH   12   . . . . . . X . . . . . . . . . . . . .
  3 C   WIDE GAP      14   . . . . . . . X . . . . . . . . . . . .
  3 C   NARROW GAP    15   . . . . . . . . X . . . . . . . . . . .
  3 C   FUEL PROCESS  18   . . . . . . . . . X . . . . . . . . . .
  3 C   CENTTEMPLIM   23   . . . . . . . . . . X . . . . . . . . .
  3 C   MOD HEAT FR.  24   . . X . . X X X . . X . . . . . . . . .
  3 C   CONTROL SPAN  25   . . . . . X . . . X . X . . . . . . . .
  2 D   BYPASS FRACT  26   . . . . . . . . . . . X . X . . . . . .
  2 D   SEP.CARY/UND  27   . . . . . . . . . . . . . . . X . . . .
  2 D   EXIT QUALITY  32   . . . . . . . . . . . . . . . X X . . .
  2 D   JET PUMP EFF  35   . . . . . . . . . . . . . . . . X . . .
  2 E   GAP H/T COEF  45   . . . . . . . . . . . . . . . . . X . .
  3 F   CORE LENGTH   48   . . . . . . . . . . . . . . . . . . X .
  3 F   LOCAL   P/A   49   . . . . . . . . . . . . . . . . . . . X
  3 F   AXIAL   P/A   50   . . . . . . . . . . . . . . . . . . . .
  3 F   RADIAL  P/A   51   . . . . . . . . . . . . . . . . . . . .
  2 D   ORIFICE PATT  29   . . . . . . . . . . . . . . . . . . . .
  3 F   TOTAL P/A     52   . . . . . . . . . . . . . . . . . . . X
  3 G   STABIL.CRIT.  56   . . . . . . . . . . . . . . . . . . . .
  4 H   RESIDENCE TM  57   . . . . . . . . . . . . . . . . . . . .
  4 H   CRUD RATE     58   . . . . . . . . . . . . . . . . . . . .
  1 I   CAP. FACTOR   59   . . . . . . . . . . . . . . . . . . . .
  1 I   REFUEL PLAN   60   . . . . . . . . . . . . . . . . . . . .
  3 C   FUEL ROD O.D  22   . . X X 8 X . . . . . . . . . . . . . .
  3 C   PELLET-CLAD   21   . 8 . . . X . . X X . . . . . . . . . .
  3 C   UO2 DIAM      19   . . . . X . . . . . . . . . . . . . . .
  2 D   HYDR. DIAM.   28   . . . X . . 8 9 9 . . . . . . . . . . .
  2 E   AVG HEAT FLX  40   . . . . . . . . . . . . . . . . . . . .
  3 C   UO2 DENSITY   16   . . . . . . . . 8 X . . . . . . . . . .
  3 C   EXPOSURE      17   . . . . . . . . . . . . . . . . . . . .
  2 E   SFC H/T COEF  38   . X . . . . . . . . . . . . . . . . . .
  2 E   MAX HEAT FLX  39   . . . . . . . . 8 . . . . . . . . . . .
  2 E   AVG SPEC POW  44   . 9 . . . . . . . X . . . . . . . 8 . .
  2 E   H/T AREA/BDL  41   . . . X . . . . . . . . . . . . . . X .
  2 E   POW DENSITY   42   . . . . . . X . . . . 8 . . . . . X . .
  1 A   MWT            2   X . . . . . . . . . . . . . . . . . . .
  1 A   STEAM FLOW     5   . X X . . . . . . . . . . . . . . . . .
  2 B   PUMPING POW    7   . X . . . . . . . . . . . . . . X 8 . .
  2 B   THERM EFFIC.   8   . X X . . . . . . . . . . . . . . . . .
  2 D   FLOW/BUNDLE   30   . . . . X . . . . . . . . . . . X . . .
  2 D   CORE PRESDRP  36   . . . X . X X X . . . . . . . . X . X .
  2 D   ORIF PRESDRP  37   . . . . . . . . . . . . . . . . X . X .
  3 F   CORE VOL.     46   . . . . . . . . . . . . . . . . . . . .
  3 F   CORE RADIUS   47   . . . . . . . . . . . . . . . . . . X .
  1 A   SEC.SYST&TUR   6   X X X . . . . . . . . . . . . . . . . .
  3 C   CHAN. THICK   13   . . . . . X 8 8 . . . . . . . . . . . .
  3 C   CLAD THICK    20   . . . . . X . . . . . . . . . . . . . .
  2 D   TOTAL FLOW    31   . . . . . . . . . . . . . . . . X . . .
  2 D   AVG VOIDS     33   . X X . . . . . . . . . . 8 9 X . . . .
  2 D   STEAM SEPFLO  34   . . . . . . . . . . . . . . . X . . . .
  2 E   MCHFR         43   . 8 8 . . . . . . . . . . . . . X . . .
  3 F   VOID COEFF    53   . X . X . . . . . . . . . . . . . . 8 .
  3 G   VOID TRANTIM  54   . . 8 . . . . . . . . . . . . . X . . .
  3 G   FUELTIMCONST  55   . . . . . . . . 8 . . . . . . X . . . .
```

Figure 7.15c

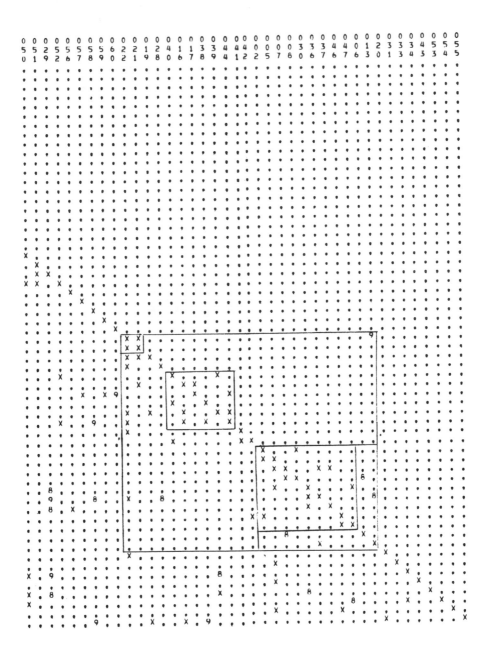

```
        BLOCK SIZE    6
     SHUNT DIAGRAM
     PRINCIPAL CIRCUIT LENGTH    6
     VARIABLE NAME                      E     V    2    3
        2 E        SFC H/T COEF        38    38         I
        2 E        AVG SPEC POW        44    44    B    E
        3 C        EXPOSURE            17    17    I
        3 C        U02 DENSITY         16    16    I    B
        2 E        MAX HEAT FLX        39    39    E    I
        2 E        AVG HEAT FLX        40    40         I
```

Figure 7.15d

```
     BLOCK SIZE    6
   SHUNT DIAGRAM SUMMARY
      2 SHUNTS

VARIABLE NAME                   F    V    NB   NE
                                                     FS   NS   NS-NB   NS-NE
   2 E       SFC H/T COEF      38   38    0    0
                                                     3    1      1      0 ———
   2 E       AVG SPEC POW      44   44    1    1
                                                     2    1      0      1
   3 C       EXPOSURE          17   17    0    0
                                                     2    1      1      1
   3 C       U02 DENSITY       16   16    1    0
                                                     2    2      1      1
   2 E       MAX HEAT FLX      39   39    0    1
                                                     3    1      1      1
   2 E       AVG HEAT FLX      40   40    0    0
                                                     3    1      1      1
```

Figure 7.15e

```
        BLOCK SIZE    9
     SHUNT DIAGRAM
     PRINCIPAL CIRCUIT LENGTH    7
     VARIABLE NAME                      E     V    0    1    3
        2 D        FLOW/BUNDLE         30    30    B    E    I
        2 D        CORE PRESDRP        36    36    37        5
        2 B        PUMPING POW          7     7    E         E
        2 B        THERM EFFIC.         8     8
        1 A        MWT                  2     2         B    B
        3 F        CORE VOL.           46    46         I    I
        3 F        CORE RADIUS         47    47         5    I
```

Figure 7.15f

```
BLOCK SIZE   9
SHUNT DIAGRAM SUMMARY
  3 SHUNTS
```

VARIABLE NAME		E	V	NB	NE	FS	NS	NS-NB	NS-NE
2 D	FLOW/BUNDLE	30	30	1	1	0	2	1	2
2 D	CORE PRESDRP	36	36	0	0	0	2	2	0
2 B	PUMPING POW	7	7	0	2	-1	0	0	0
2 B	THERM EFFIC.	8	8	0	0	-1	0	0	0
1 A	MWT	2	2	2	0	1	2	0	2
3 F	CORE VOL.	46	46	0	0	1	2	2	2
3 F	CORE RADIUS	47	47	0	0	1	2	2	1

Figure 7.15g

```
INSERT TEARS AT LEVEL 5
ORDER  60
HIGHEST LEVEL MARK TO BE CONSIDERED   9
LARGEST BLOCK FOR WHICH SHUNT DIAGRAM IS NOT PRINTED   2
LARGEST INDEX SHUNT TO BE CONSIDERED   60
CHANGES    ROW COL SEN
           44  38   5
           44  16   5
            8   7   5
           22  21   5
```

```
LEVEL  9
  1  1  1  1  1  1  1  1  1  1  1  1  1  1  1  1  1  1  1
  1  1  1  1  1  1  1  1  1 23  1  1  1  1  1  1  1  1
```

```
LEVEL  8
  1  1  1  1  1  1  1  1  1  1  1  1  1  1  1  1  1  1  1
  1  1  1  1  1  1  1  1  1  2  1  1  6  1  1 11  1  1  1  1
  1  1  1  1
```

```
LEVEL  5
  1  1  1  1  1  1  1  1  1  1  1  1  1  1  1  1  1  1  1
  1  1  1  1  1  1  1  1  1  2  1  1  6  1  1  9  1  1  1  1
  1  1  1  1  1  1
```

```
LEVEL  0
  1  1  1  1  1  1  1  1  1  1  1  1  1  1  1  1  1  1  1
  1  1  1  1  1  1  1  1  1  1  1  1  1  1  1  1  1  1  1  1
  1  1  1  1  1  1  1  1  1  1  1  1  1  1  1  1  1  1  1  1
```

Figure 7.15h

INSERT TEARS AT LEVEL 5
PAGE 1/ 1

AFTER PARTITIONING

			Row
1	A	MWE	1
1	A	PRESSURE-SAT	3
1	A	FEEDWAT-TEMP	4
3	C	MOD/FUEL VOL	9
3	C	RODS/BUNDLE	10
3	C	CLAD THK/DIA	11
3	C	CONT. PITCH	12
3	C	WIDE GAP	14
3	C	NARROW GAP	15
3	C	FUEL PROCESS	18
3	C	CENTTEMPLIM	23
3	C	MOD HEAT FR.	24
3	C	CONTROL SPAN	25
2	D	BYPASS FRACT	26
2	D	SEP.CARY/UND	27
2	D	EXIT QUALITY	32
2	D	JET PUMP EFF	35
2	E	GAP H/T COEF	45
3	F	CORE LENGTH	48
3	F	LOCAL P/A	49
3	F	AXIAL P/A	50
3	F	RADIAL P/A	51
2	D	ORIFICE PATT	29
3	F	TOTAL P/A	52
3	G	STABIL.CRIT.	56
4	H	RESIDENCE TM	57
4	H	CRUD RATE	58
1	I	CAP. FACTOR	59
1	I	REFUEL PLAN	60
3	C	FUEL ROD O.D	22
3	C	PELLET-CLAD	21
3	C	UO2 DIAM	19
2	D	HYDR. DIAM.	28
2	E	AVG SPEC POW	44
3	C	EXPOSURE	17
3	C	UO2 DENSITY	16
2	E	MAX HEAT FLX	39
2	E	AVG HEAT FLX	40
2	E	SFC H/T COEF	38
2	E	H/T AREA/BDL	41
2	E	POW DENSITY	42
2	B	THERM EFFIC.	8
1	A	MWT	2
1	A	STEAM FLOW	5
3	F	CORE VOL.	46
3	F	CORE RADIUS	47
2	D	FLOW/BUNDLE	30
2	D	CORE PRESDRP	36
2	D	ORIF PRFSDRP	37
2	B	PUMPING POW	7
1	A	SEC.SYST&TUR	6
3	C	CHAN. THICK	13
3	C	CLAD THICK	20
2	D	TOTAL FLOW	31
2	D	AVG VOIDS	33
2	D	STEAM SEPFLO	34
2	E	MCHFR	43
3	F	VOID COEFF	53
3	G	VOID TRANTIM	54
3	G	FUELTIMCONST	55

Figure 7.15i

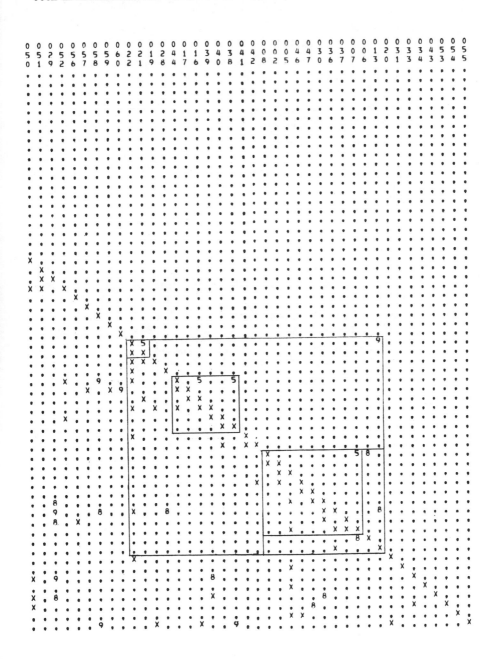

determination of the Fuel Rod O.D. Thus we tear the mark showing Pellet-Clad preceding Fuel Rod O.D.

Next let us consider the block of size 9. We begin by examining the Shunt Diagram Summary in Figure 7.15g. We first look in the Number of Shunts (NS) column and note that there are two arcs in the principal circuit with no shunts, either (8,7) which is the effect of Pumping Pow on Therm Effic., or (2,8) which is the effect of Therm Effic. on Megawatts Thermal. The -1 in column FS merely implies that there is no parallel shunt. The pumping power makes only a small contribution to the thermal efficiency. However, the thermal efficiency is a direct multiplier in computing megawatts thermal so that a 1% error in the one will yield a 1% error in the other. Therefore, of these two we choose to tear the effect of Pumping Pow on Therm Effic., (8,7). We do not notice any other tears in the principal circuit which look any better. Now we can go to Figure 7.15f to apply this choice to the shunt diagram. Drawing a horizontal line between rows 7 and 8, we see that the line does not intersect any shunt parallel to the arc (8,7) in the principal circuit, as we had expected from looking at the shunt diagram summary.

Now we consider tearing the block of size 6. In the Shunt Diagram Summary in Figure 7.15e, we look down the NS (Number of Shunts) column and see that all the arcs in the principal circuit have 1 or 2 shunts. The highest index First Shunt (FS) is 3, and there are several of those. We might consider tearing one of the arcs where NS is 1 and FS is 3. But let us look further for other possibilities.

The lines in the Shunt Diagram Summary represent vertices in the principal circuit and the half-lines represent the arcs between these adjacent vertices in the principal circuit. We look at the Number of B's (NB) and subtract it from the Number of Shunts (NS) on the next half-line to obtain the number of shunts which are not due to arcs exiting from this vertex. This is printed in a column under NS-NB. It gives an indication of how many additional shunts we will have to consider after we have torn the arcs exiting this vertex. Tearing arcs exiting a vertex means tearing marks in the column corresponding to that vertex in the precedence matrix. For example, we see that Ave Spec Pow (44) has 1 for NB and a 1 for NS on the following half-line. NS-NB on the half line is 0.

Let us now look at the Shunt Diagram in Figure 7.15d. Drawing a line between Ave Spec Pow and Exposure intersects the first shunt just below the E and misses the other shunt completely. Thus by tearing two arcs—(17,44) in the principal circuit and (39,44) corresponding to the B in the shunt, both exiting 44 and thus corresponding to marks in column 44 in the precedence matrix—we would expect to break all the circuits in that block.

Similarly, we can look at the number of E's (NE) and subtract it from the Number of Shunts (NS) on the previous half-line to obtain the number of shunts which are not due to arcs entering this vertex of the principal circuit. This number appears in the column labeled NS-NE. (Note that the half-line previous to the first line is the last half-line because the diagram is circular.) This number gives an indication of how many additional tears have to be considered after we have torn the arcs entering this vertex. The arcs entering a vertex correspond to marks in the row of that vertex in the precedence matrix. For example, we see that Ave Spec Pow (44) has a 1 in NE and a 1 in NS in the preceding half-row. Drawing a line on the shunt diagram in Figure 7.15d between Spec H/T Coef and Ave Spec Pow intersects the second shunt just before the E and misses the other shunt completely. Thus be tearing the two arcs—(44,38) in the principal circuit and (44,16) corresponding to the E in the shunt, both arcs entering vertex 44 and thus corresponding to marks in row 44 of the precedence matrix— we would again expect to break all the circuits in the block.

Choosing between these possibilities we decide to tear (44,38) and (44,16) on the grounds that this set of tears is not unreasonable from the engineering point of view and also represents a set of tears in one row, thus minimizing the number of variables directly affected by the estimates. Figures 7.15h and 7.15i show the results of this tearing. Part i now represents a good design structure matrix which depicts the design procedure to all of the engineers involved.

Figure 7.16 shows the IMPACT list generated by the Inversion of Lists program. Note that the tasks have been renumbered by the Renumber and Sort program to put them in the order obtained by TERABL. Next to each predecessor or successor is the level number. An asterisk next to a predecessor or successor implies that this task and that predecessor or successor are involved in a common block.

Now let us consider how the ASPECT program can be used from a time-share terminal to trace through this matrix to determine what tasks will be affected if certain tasks are changed; see Figure 7.17.

The design structure matrix with the tasks renumbered according to the order they appear in this matrix are read onto a disk file. This file may be interrogated from a time-share terminal using the ASPECT program. The user will have a notebook which gives the external number for each task. The use of a notebook and external number avoids problems due to ambiguity in recognizing the spelling or abbreviation of the task names that might be typed by the user. Let us suppose that he wishes to consider what would be affected if he were to change the system pressure and consequent saturated temperature. The notebook shows that this is task 2. The terminal asks the user to "Type External Task Number or 0" and prints

```
        ITEM
                  PREDECESSORS

    1    1 A         MWE


    2    1 A          MWT
                1    1 A          MWE
                8    2 B          THERM EFFIC.              0
                                                           0  *
    3    1 A          PRESSURE-SAT






    4    1 A          FEEDWAT-TEMP





    5    1 A          STEAM FLOW
                2    1 A          MWT                       0
                3    1 A          PRESSURE-SAT              0
                4    1 A          FEEDWAT-TEMP              0



    6    1 A          SEC.SYST&TUR
                1    1 A          MWE                       0
                3    1 A          PRESSURE-SAT              0
                4    1 A          FEEDWAT-TEMP              0
                7    2 B          PUMPING POW               8  *

    7    2 B          PUMPING POW
                3    1 A          PRESSURE-SAT              0
                5    1 A          STEAM FLOW                0
               32    2 D          EXIT QUALITY              0  *
               35    2 D          JET PUMP EFF              8  *
               36    2 D          CORE PRESDRP              0  *
               37    2 D          ORIF PRESDRP              0  *

    8    2 B          THERM EFFIC.
                3    1 A          PRESSURE-SAT              0
                4    1 A          FEEDWAT-TEMP              0
                6    1 A          SEC.SYST&TUR              8
                7    2 B          PUMPING POW               0

    9    3 C          MOD/FUEL VOL
```

196

Figure 7.16
IMPACT List

2	1 A	MWT	0
6	1 A	SEC,SYST&TUR	0
5	1 A	STEAM FLOW	0
46	3 F	CORE VOL.	0
5	1 A	STEAM FLOW	0
6	1 A	SEC,SYST&TUR	0
7	2 B	PUMPING POW	0
8	2 B	THERM EFFIC.	0
21	3 C	PELLET-CLAD	8
33	2 D	AVG VOIDS	0
38	2 E	SFC H/T COEF	0
43	2 E	MCHFR	8
44	2 E	AVG SPEC POW	9
53	3 F	VOID COEFF	0
5	1 A	STEAM FLOW	0
6	1 A	SEC,SYST&TUR	0
8	2 B	THERM EFFIC.	0
33	2 D	AVG VOIDS	0
43	2 E	MCHFR	8
54	3 G	VOID TRANTIM	8
7	2 B	PUMPING POW	0
30	2 D	FLOW/BUNDLE	0
31	2 D	TOTAL FLOW	0
34	2 D	STEAM SEPFLO	0
54	3 G	VOID TRANTIM	0
8	2 B	THERM EFFIC.	8
6	1 A	SEC,SYST&TUR	8 *
8	2 B	THERM EFFIC.	0
2	1 A	MWT	0 *
22	3 C	FUEL ROD O.D	0
24	3 C	MOD HEAT FR.	0
53	3 F	VOID COEFF	0

197

```
      ITEM
                      PREDECESSORS

   10    3 C          RODS/BUNDLE



   11    3 C          CLAD THK/DIA



   12    3 C          CONT. PITCH




   13    3 C          CHAN. THICK
                12    3 C          CONT. PITCH              0
                14    3 C          WIDE GAP                 8 ★
                15    3 C          NARROW GAP               8 ★
                36    2 D          CORE PRESDRP             0 ★

   14    3 C          WIDE GAP



   15    3 C          NARROW GAP




   16    3 C          UO2 DENSITY
                17    3 C          EXPOSURE                 0 ★
                18    3 C          FUEL PROCESS             8 ★
                21    3 C          PELLET-CLAD              0 ★
                23    3 C          CENTTEMPLIM              0 ★

   17    3 C          EXPOSURE
                57    4 H          RESIDENCE TM             0 ★
                44    2 E          AVG SPEC POW             0 ★
                59    1 I          CAP. FACTOR              0 ★
                60    1 I          REFUEL PLAN              9 ★

   18    3 C          FUEL PROCESS
```

Figure 7.16
(continued)

SUCCESSORS

22	3 C	FUEL ROD O.D	0	
28	2 D	HYDR. DIAM.	0	
36	2 D	CORE PRESDRP	0	
41	2 E	H/T AREA/BDL	0	
19	3 C	UO2 DIAM	0	
20	3 C	CLAD THICK	0	
21	3 C	PELLET-CLAD	0	
22	3 C	FUEL ROD O.D	8	
13	3 C	CHAN. THICK	0	
22	3 C	FUEL ROD O.D	0	
24	3 C	MOD HEAT FR.	0	
25	3 C	CONTROL SPAN	0	
28	2 D	HYDR. DIAM.	8	
30	2 D	FLOW/BUNDLE	0	
36	2 D	CORE PRESDRP	0	
42	2 E	POW DENSITY	0	
22	3 C	FUEL ROD O.D	9	
36	2 D	CORE PRESDRP	8	
13	3 C	CHAN. THICK	8	
24	3 C	MOD HEAT FR.	0	
28	2 D	HYDR. DIAM.	9	
36	2 D	CORE PRESDRP	0	
13	3 C	CHAN. THICK	8	
24	3 C	MOD HEAT FR.	0	
28	2 D	HYDR. DIAM.	9	
36	2 D	CORE PRESDRP	0	
39	2 E	MAX HEAT FLX	0	
44	2 E	AVG SPEC POW	0	
55	3 G	FUELTIMCONST	0	
16	3 C	UO2 DENSITY	0	*
16	3 C	UO2 DENSITY	8	
21	3 C	PELLET-CLAD	0	

```
        ITEM
                    PREDECESSORS

 19  3 C          UO2 DIAM
            11   3 C          CLAD THK/DIA          0
            21   3 C          PELLET-CLAD           0  *
            22   3 C          FUEL ROD O.D          0  *

 20  3 C          CLAD THICK
            11   3 C          CLAD THK/DIA          0
            22   3 C          FUEL ROD O.D          0  *

 21  3 C          PELLET-CLAD
             3   1 A          PRESSURE-SAT          8
            11   3 C          CLAD THK/DIA          0
            18   3 C          FUEL PROCESS          0
            22   3 C          FUEL ROD O.D          0  *
            23   3 C          CENTTEMPLIM           0  *

 22  3 C          FUEL ROD O.D
             9   3 C          MOD/FUEL VOL          0
            10   3 C          RODS/BUNDLE           0
            11   3 C          CLAD THK/DIA          8
            12   3 C          CONT. PITCH           0
            13   3 C          CHAN. THICK           9
            21   3 C          PELLET-CLAD           0


 23  3 C              CENTTEMPLIM




 24  3 C          MOD HEAT FR.
             9   3 C          MOD/FUEL VOL          0
            12   3 C          CONT. PITCH           0
            14   3 C          WIDE GAP              0
            15   3 C          NARROW GAP            0

 25  3 C          CONTROL SPAN
            12   3 C          CONT. PITCH           0

 26  2 D          BYPASS FRACT
            24   3 C          MOD HEAT FR.          0

 27  2 D          SEP.CARY/UND


 28  2 D          HYDR. DIAM.
            10   3 C          RODS/BUNDLE           0
            12   3 C          CONT. PITCH           8
            14   3 C          WIDE GAP              9
            15   3 C          NARROW GAP            9
            22   3 C          FUEL ROD O.D          0
```

Figure 7.16
(continued)

SUCCESSORS

39	2	E	MAX HEAT FLX	0	
55	3	G	FUELTIMCONST	0	
55	3	G	FUELTIMCONST	0	
16	3	C	UO2 DENSITY	0	*
19	3	C	UO2 DIAM	0	*
22	3	C	FUEL ROD O.D	0	
19	3	C	UO2 DIAM	0	*
20	3	C	CLAD THICK	0	*
21	3	C	PELLET-CLAD	0	*
28	2	D	HYDR. DIAM.	0	
36	2	D	CORE PRESDRP	0	
39	2	E	MAX HEAT FLX	0	
41	2	E	H/T AREA/BDL	0	
44	2	E	AVG SPEC POW	0	
16	3	C	UO2 DENSITY	0	
21	3	C	PELLET-CLAD	0	
44	2	E	AVG SPEC POW	0	
55	3	G	FUELTIMCONST	8	
26	2	D	BYPASS FRACT	0	
39	2	E	MAX HEAT FLX	8	
42	2	E	POW DENSITY	8	
33	2	D	AVG VOIDS	8	
33	2	D	AVG VOIDS	9	
36	2	D	CORE PRESDRP	8	

```
         ITEM
                      PREDECESSORS

29   2 D          ORIFICE PATT
         51    3 F          RADIAL   P/A              0  *




30   2 D          FLOW/BUNDLE
          5    1 A          STEAM FLOW               0
         12    3 C          CONT. PITCH              0
         29    2 D          ORIFICE PATT             8
         47    3 F          CORE RADIUS              0  *
         32    2 D          EXIT QUALITY             0  *

31   2 D          TOTAL FLOW
          5    1 A          STEAM FLOW               0
         32    2 D          EXIT QUALITY             0  *

32   2 D          EXIT QUALITY




33   2 D          AVG VOIDS
          3    1 A          PRESSURE-SAT             0
          4    1 A          FEEDWAT-TEMP             0
         26    2 D          BYPASS FRACT             8
         27    2 D          SEP.CARY/UND             9
         50    3 F          AXIAL    P/A             0  *
         29    2 D          ORIFICE PATT             9
         39    2 E          MAX HEAT FLX             8  *
         32    2 D          EXIT QUALITY             0

34   2 D          STEAM SEPFLO
          5    1 A          STEAM FLOW               0
         32    2 D          EXIT QUALITY             0

35   2 D          JET PUMP EFF


36   2 D          CORE PRESDRP
         10    3 C          RODS/BUNDLE              0
         12    3 C          CONT. PITCH              0
         13    3 C          CHAN. THICK              8
         14    3 C          WIDE GAP                 0
         15    3 C          NARROW GAP               0
         22    3 C          FUEL ROD O.D             0
         28    2 D          HYDR. DIAM.              8
         29    2 D          ORIFICE PATT             9
         48    3 F          CORE LENGTH              0  *
         30    2 D          FLOW/BUNDLE              0
         32    2 D          EXIT QUALITY             0
202      58    4 H          CRUD RATE                8  *
```

SUCCESSORS

30	2	D	FLOW/BUNDLE	8
33	2	D	AVG VOIDS	9
36	2	D	CORE PRESDRP	9
37	2	D	ORIF PRESDRP	8
43	2	E	MCHFR	8

36	2	D	CORE PRESDRP	0
37	2	D	ORIF PRESDRP	0
43	2	E	MCHFR	8

7	2	B	PUMPING POW	0
30	2	D	FLOW/BUNDLE	0
31	2	D	TOTAL FLOW	0
33	2	D	AVG VOIDS	0
34	2	D	STEAM SEPFLO	0
36	2	D	CORE PRESDRP	0
43	2	E	MCHFR	0
54	3	G	VOID TRANTIM	0

53	3	F	VOID COEFF	0

7	2	B	PUMPING POW	8

7	2	B	PUMPING POW	0 *
13	3	C	CHAN, THICK	0 *

Figure 7.16
(continued)

PREDECESSORS

```
37   2 D        ORIF PRESDRP
        29   2 D        ORIFICE PATT         8
        30   2 D        FLOW/BUNDLE          0
        56   3 G        STABIL.CRIT.         0 *

38   2 E        SFC H/T COEF
         3   1 A        PRESSURE-SAT         0
        40   2 E        AVG HEAT FLX         0 *

39   2 E        MAX HEAT FLX
        16   3 C        UO2 DENSITY          0
        19   3 C        UO2 DIAM             0
        22   3 C        FUEL ROD O.D         0
        24   3 C        MOD HEAT FR.         8
        44   2 E        AVG SPEC POW         0 *

40   2 E        AVG HEAT FLX
        52   3 F        TOTAL P/A            0 *
        39   2 E        MAX HEAT FLX         0

41   2 E        H/T AREA/BDL
        10   3 C        RODS/BUNDLE          0
        48   3 F        CORE LENGTH          0 *
        22   3 C        FUEL ROD O.D         0

42   2 E        POW DENSITY
        12   3 C        CONT. PITCH          0
        24   3 C        MOD HEAT FR.         8
        40   2 E        AVG HEAT FLX         0
        41   2 E        H/T AREA/BDL         0
        48   3 F        CORE LENGTH          0 *

43   2 E        MCHFR
         3   1 A        PRESSURE-SAT         8
         4   1 A        FEEDWAT-TEMP         8
        50   3 F        AXIAL   P/A          0 *
        29   2 D        ORIFICE PATT         8
        39   2 E        MAX HEAT FLX         0
        30   2 D        FLOW/BUNDLE          8
        32   2 D        EXIT QUALITY         0

44   2 E        AVG SPEC POW
         3   1 A        PRESSURE-SAT         9
        16   3 C        UO2 DENSITY          0
        22   3 C        FUEL ROD O.D         0
        23   3 C        CENTTEMPLIM          0
        38   2 E        SFC H/T COEF         0
        52   3 F        TOTAL P/A            0 *
        58   4 H        CRUD RATE            9 *
        45   2 E        GAP H/T COEF         8 *

45   2 E        GAP H/T COEF

46   3 F        CORE VOL.
         2   1 A        MWT                  0
        42   2 E        POW DENSITY          0
```

Figure 7.16
(continued)

SUCCESSORS

7	2 B	PUMPING POW	0 *

| 44 | 2 E | AVG SPEC POW | 0 |
| 55 | 3 G | FUELTIMCONST | 9 |

33	2 D	AVG VOIDS	8 *
40	2 E	AVG HEAT FLX	0
43	2 E	MCHFR	0

| 38 | 2 E | SFC H/T COEF | 0 * |
| 42 | 2 E | POW DENSITY | 0 |

| 42 | 2 E | POW DENSITY | 0 |

| 46 | 3 F | CORE VOL. | 0 |

| 17 | 3 C | EXPOSURE | 0 * |
| 39 | 2 E | MAX HEAT FLX | 0 * |

| 44 | 2 E | AVG SPEC POW | 8 |
| 55 | 3 G | FUELTIMCONST | 0 |

| 47 | 3 F | CORE RADIUS | 0 |
| 54 | 3 G | VOID TRANTIM | 0 |

205

ITEM			PREDECESSORS				
47	3 F	CORE RADIUS					
			46	3 F	CORE VOL.	0	
			48	3 F	CORE LENGTH	0	*
48	3 F	CORE LENGTH					
49	3 F	LOCAL P/A					
50	3 F	AXIAL P/A					
51	3 F	RADIAL P/A					
52	3 F	TOTAL P/A					
			49	3 F	LOCAL P/A	0	
			50	3 F	AXIAL P/A	0	
			51	3 F	RADIAL P/A	0	
53	3 F	VOID COEFF					
			3	1 A	PRESSURE-SAT	0	
			9	3 C	MOD/FUEL VOL	0	
			50	3 F	AXIAL P/A	0	
			48	3 F	CORE LENGTH	8	
			47	3 F	CORE RADIUS	8	
			33	2 D	AVG VOIDS	0	
54	3 G	VOID TRANTIM					
			4	1 A	FEEDWAT-TEMP	8	
			5	1 A	STEAM FLOW	0	
			46	3 F	CORE VOL.	0	
			32	2 D	EXIT QUALITY	0	
55	3 G	FUELTIMCONST					
			16	3 C	UO2 DENSITY	0	
			19	3 C	UO2 DIAM	0	
			20	3 C	CLAD THICK	0	
			23	3 C	CENTTEMPLIM	8	
			38	2 E	SFC H/T COEF	9	
			58	4 H	CRUD RATE	9	*
			45	2 E	GAP H/T COEF	0	
56	3 G	STABIL.CRIT.					
57	4 H	RESIDENCE TM					

Figure 7.16
(continued)

SUCCESSORS

30	2	D	FLOW/BUNDLE	0	*
53	3	F	VOID COEFF	8	
36	2	D	CORE PRESDRP	0	
41	2	E	H/T AREA/BDL	0	
42	2	E	POW DENSITY	0	
47	3	F	CORE RADIUS	0	
53	3	F	VOID COEFF	8	
52	3	F	TOTAL P/A	0	
33	2	D	AVG VOIDS	0	
43	2	E	MCHFR	0	
52	3	F	TOTAL P/A	0	
53	3	F	VOID COEFF	0	
29	2	D	ORIFICE PATT	0	
52	3	F	TOTAL P/A	0	
40	2	E	AVG HEAT FLX	0	*
44	2	E	AVG SPEC POW	0	*
37	2	D	ORIF PRESDRP	0	
17	3	C	EXPOSURE	0	

```
    ITEM
                PREDECESSORS
58    4  H      CRUD RATE



59    1  I      CAP. FACTOR


60    1  I      REFUEL PLAN
```

Figure 7.16
(continued)

" = " to show it is ready to receive input. He types "2." The terminal prints the 2 and the name it has for that task so the user can confirm that he has typed the correct number. Each time the terminal prints this message the user may type the number of another task or type 0 to terminate this input. The terminal then prints "Command?". If the user responds with "SUC" the terminal will print all the immediate successors for each task which appeared in the last step. At this point it would cause all the immediate successors of "Pressure-Sat" to be printed. This would be equivalent to going down the "Pressure-Sat" column of the design structure matrix to determine all other tasks which use this task directly. The next time the "SUC" command is given the successors of each of these tasks will be printed.

For each successor the terminal prints the external number, name, level of the mark representing the sensitivity of the successor, and position of the successor relative to the task it succeeds, that is, the difference between the internal numbers. If this relative position is negative it implies that the corresponding mark in the matrix is above the diagonal. If this number is large and negative it implies that it occurred in a large block in the original design structure matrix.

If a task has occurred before, the word "Repeat" is printed and this

SUCCESSORS

36	2	D	CORE PRESDRP	8
44	2	E	AVG SPEC POW	9
55	3	G	FUELTIMCONST	9
፡7	3	C	EXPOSURE	0
17	3	C	EXPOSURE	9

task is not followed again. If no further successors exist within the system for this task, the message "No Successor" is printed.

The user can review the successors to determine whether they are likely to change. He can reject tasks which appeared in the last set with the "REJ" command, or he can select the tasks in the last set to be retained by the "SEL" command. This allows the user to interact to prevent many meaningless tasks from being generated as successors to some task which in the user's judgment will not be affected sufficiently to warrant change.

Once the successor tasks are generated it is possible to retrieve and print data associated with these tasks. This data might be the durations used to schedule the work. In this example the data files contained the names of the responsible engineers and the estimated man-hours for these tasks. The command is composed of letters as follows: P stands for Print, T stands for These (meaning only the tasks appearing in the last step) while A stands for All (meaning all of the tasks which have been generated), E stands for responsible engineers, and D stands for Data. Thus, to print the names of the responsible engineers for all of the tasks which have been generated the user types "PAE."

The "INP" command clears the list of tasks generated and begins a new list by requesting new input.

```
RUN ASPECT

TYPE EXTERNAL TASK NUMBER OR 0
= 2
   2    1 A          PRESSURE-SAT
TYPE EXTERNAL TASK NUMBER OR 0
= 0

COMMAND?
= SUC

SUCCESSORS OF       2    1 A          PRESSURE-SAT
   NO.      TASK                              SEN  REL
   31     3 C         PELLET-CLAD              8   29
   34     2 E         AVG SPEC POW             9   32
   39     2 E         SFC H/T COEF             0   37
   42     2 B         THERM EFFIC.             0   40
   44     1 A         STEAM FLOW               0   42
   50     2 B         PUMPING POW              0   48
   51     1 A         SEC.SYST&TUR             0   49
   55     2 D         AVG VOIDS                0   53
   57     2 E         MCHFR                    8   55
   58     3 F         VOID COEF                0   56

COMMAND?
= SUC

SUCCESSORS OF      31    3 C          PELLET-CLAD
   NO.       TASK                             SEN  REL
   30     3 C         FUEL ROD O.D.            0   -1
   32     3 C         UO2 DIAM                 0    1
   36     3 C         UO2 DENSITY              0    5

SUCCESSORS OF      34    2 E          AVG SPEC POW
   NO.       TASK                             SEN  REL
   35     3 C         EXPOSURE                 0    1
   37     2 E         MAX HEAT FLX             0    3

SUCCESSORS OF      39    2 E          SFC H/T COEF
   NO.       TASK                             SEN  REL
   34     2 E         AVG SPEC POW             0   -5   REPEAT
   60     3 G         FUELTIMCONST             9   21

SUCCESSORS OF      42    2 B          THERM EFFIC.
   NO.       TASK                             SEN  REL
   43     1 A         MWT                      0    1

SUCCESSORS OF      44    1 A          STEAM FLOW
   NO.       TASK                             SEN  REL
   47     2 D         FLOW/BUNDLE              0    3
   50     2 B         PUMPING POW              0    6   REPEAT
   54     2 D         TOTAL FLOW               0   10
   56     2 D         STEAM SEPFLO             0   12
   59     3 G         VOID TRANTIM             0   15
```

Figure 7.17
ASPECT Time Share Run

```
SUCCESSORS OF     50    2 B      PUMPING POW
  NO.      TASK                              SEN REL
   42     2 B        THERM EFFIC.              0  -8   REPEAT
   51     1 A        SEC.SYST&TUR             8   1   REPEAT

SUCCESSORS OF     51    1 A      SEC.SYST&TUR
  NO.      TASK                              SEN REL
   42     2 B        THERM EFFIC.             8  -9   REPEAT

SUCCESSORS OF     55    2 D      AVG VOIDS
  NO.      TASK                              SEN REL
   58     3 F        VOID COEFF               0   3   REPEAT

SUCCESSORS OF     57    2 E      MCHFR
  NO.      TASK                              SEN REL
      NO SUCCESSOR

SUCCESSORS OF     58    3 F      VOID COEFF
  NO.      TASK                              SEN REL
      NO SUCCESSOR

COMMAND?
= REJ
REJECT - TYPE NUMBER OR 0
= 30
= 42
= 0

COMMAND?
= SUC

SUCCESSORS OF     32    3 C      UO2 DIAM
  NO.      TASK                              SEN REL
   37     2 E        MAX HEAT FLX             0   5   REPEAT
   60     3 G        FUELTIMCONST             0  28   REPEAT

SUCCESSORS OF     36    3 C      UO2 DENSITY
  NO.      TASK                              SEN REL
   34     2 E        AVG SPEC POW             0  -2   REPEAT
   37     2 E        MAX HEAT FLX             0   1   REPEAT
   60     3 G        FUELTIMCONST             0  24   REPEAT

SUCCESSORS OF     35    3 C      EXPOSURE
  NO.      TASK                              SEN REL
   36     3 C        UO2 DENSITY              0   1   REPEAT

SUCCESSORS OF     37    2 E      MAX HEAT FLX
  NO.      TASK                              SEN REL
   38     2 E        AVG HEAT FLX             0   1
   55     2 D        AVG VOIDS                8  18   REPEAT
   57     2 E        MCHFR                    0  20   REPEAT
```

Figure 7.17
(continued)

```
SUCCESSORS OF      60   3 G      FUELTIMCONST
  NO.      TASK                              SEN REL
      NO SUCCESSOR

SUCCESSORS OF      43   1 A      MWT
  NO.      TASK                              SEN REL
   44    1 A           STEAM FLOW             0    1    REPEAT
   45    3 F           CORE VOL.              0    2

SUCCESSORS OF      47   2 D      FLOW/BUNDLE
  NO.      TASK                              SEN REL
   48    2 D           CORE PRESDRP           0    1
   49    2 D           ORIF PRESDRP           0    2
   57    2 E           MCHFR                  8   10    REPEAT

SUCCESSORS OF      54   2 D      TOTAL FLOW
  NO.      TASK                              SEN REL
      NO SUCCESSOR

SUCCESSORS OF      56   2 D      STEAM SEPFLO
  NO.      TASK                              SEN REL
      NO SUCCESSOR

SUCCESSORS OF      59   3 G      VOID TRANTIM
  NO.      TASK                              SEN REL
      NO SUCCESSOR
```

Figure 7.17
(continued)

```
COMMAND?
= PAE
   2    1 A      PRESSURE-SAT
   THERMO-HYDR. ENG.
  31    3 C      PELLET-CLAD
   FUEL DESIGNER-MECH
  34    2 E      AVG SPEC POW
   CORE DESIGNER
  39    2 E      SFC H/T COEF
   THERMO-HYDR. ENG.
  42    2 B      THERM EFFIC.
   CORE DESIGNER
   BALANCE OF PLANT
  44    1 A      STEAM FLOW
   THERMO-HYDR. ENG.
  50    2 B      PUMPING POW
   CORE DESIGNER
   BALANCE OF PLANT
  51    1 A      SEC.SYST&TUR
   BALANCE OF PLANT
  55    2 D      AVG VOIDS
   THERMO-HYDR. ENG.
  57    2 E      MCHFR
   CRIT.& STANDARDS
  58    3 F      VOID COEFF
   NUCLEAR ENG
  32    3 C      UO2 DIAM
   FUEL DESIGNER-MECH
   NUCLEAR ENG.
  36    3 C      UO2 DENSITY
   MANUFACTURING
  35    3 C      EXPOSURE
   NUCLEAR ENG
  37    2 E      MAX HEAT FLX
   FUEL DESIGNER-MECH
   NUCLEAR ENG.
  60    3 G      FUELTIMCONST
   FUEL DESIGNER-MECH
  43    1 A      MWT
   CORE DESIGNER
  47    2 D      FLOW/BUNDLE
   THERMO-HYDR. ENG
  54    2 D      TOTAL FLOW
   CORE DESIGNER
   THERMO-HYDR. ENG
  56    2 D      STEAM SEPFLO
   THERMO-HYDR. ENG
  59    3 G      VOID TRANTIM
   FUEL DESIGNER-MECH
   THERMO-HYDR. ENG
```

Figure 7.17
(continued)

```
 38    2 E      AVG HEAT FLX
    FUEL DESIGNER-MECH
    NUCLEAR ENG.
 45    3 F      CORE VOL.
    CORE DESIGNER
 48    2 D      CORE PRESDRP
    CORE DESIGNER
    THERMO-HYDR. ENG
 49    2 D      ORIF PRESDRP
    THERMO-HYDR. ENG
COMMAND?
= PAD
  2    1 A      PRESSURE-SAT
    18 MAN-HR
 31    3 C      PELLET-CLAD
    60 MAN-HR
 34    2 E      AVG SPEC POW
    70 MAN-HR
 39    2 E      SFC H/T COEF
    35 MAN-HR
 42    2 B      THERM EFFIC.
    65 MAN-HR
 44    1 A      STEAM FLOW
    35 MAN-HR
 50    2 B      PUMPING POW
    35 MAN-HR
 51    1 A      SEC.SYST&TUR
    65 MAN-HR
 55    2 D      AVG VOIDS
    35 MAN-HR
 57    2 E      MCHFR
    25 MAN-HR
 58    3 F      VOID COEF
    24 MAN-HR
 32    3 C      UO2 DIAM
    20 MAN-HR
 36    3 C      UO2 DENSITY
    15 MAN-HR
 35    3 C      EXPOSURE
    20 MAN-HR
 37    2 E      MAX HEAT FLX
    35 MAN-HR
 60    3 G      FUELTIMCONST
    55 MAN-HR
 43    1 A      MWT
    15 MAN-HR
 47    2 D      FOLW/BUNDLE
    45 MAN-HR
 54    2 D      TOTAL FLOW
    15 MAN-HR
 56    2 D      STEAM SEPFLO
    45 MAN-HR
```

Figure 7.17
(continued)

```
59      3 G      VOID TRANTIM
     18 MAN-HR
38      2 E      AVG HEAT FLX
    110 MAN-HR
45      3 F      CORE VOL.
     18 MAN-HR
48      2 D      CORE PRESDRP
     85 MAN-HR
49      2 D      ORIF PRESDRP
     15 MAN-HR

COMMAND?
= STOP
```

Figure 7.17
(continued)

8 The Analysis of Models

8.1 The Art of Modeling

We shall consider in this chapter how partitioning and tearing can be used in the development of models. To put this into context we first define what we mean by *models* and discuss how they are developed.

We consider a *model* to be a symbolically formulated system intended to approximately represent some aspect of the behavior of another system. The model is usually described by a state and a procedure. The *state* is a sufficiently complete description of the conditions of the parts of the system at any point in time that the procedure operating on the state can determine the state at another point in time. The *procedure* incorporates a knowledge of how each of the parts is affected by other parts of the system. *Simulation* is the successive application of the procedure to generate states at different points in time.

The development of models can be considered in the following six steps:

1. *Formulation:* definition of the state and the procedure operating on that state, which describes the model

2. *Estimation:* use of data from the system to evaluate the free parameters of the model

3. *Simulation:* generation of the behavior to be predicted by the model through the repetitive application of the procedure to the successive states
4. *Validation:* comparison of the behavior simulated by the model with the behavior of that aspect of the system the model is intended to emulate in order to measure the goodness of the model
5. *Diagnosis:* tracing the difference between the behavior of the simulation and the behavior of the emulated aspect of the system in order to determine how the model can be improved
6. *Refinement:* modifying the model to improve its validity

Usually a model is a simple system used to represent some aspect of a more complex system in order to understand that system. A model represents a hypothesis that the simulation of the model will result in behavior similar to the behavior of that aspect of the system one is attempting to understand. The model may represent behavior in the small, e.g., the behavior at the next point in time, or the behavior of individual parts as they are affected by certain other parts. Simulation may then be used to generate the behavior in the large, e.g., how the system behaves over a longer period of time, or how certain effects propagate through the whole system.

The development of models through the processes of validation, diagnosis, and refinement depends upon some measure for comparing the behavior of the model with the system it is to emulate. The process of model development is very similar to the process of adaptation of systems described by Alexander (1964) and Ashby (1960). They point out that systems will adapt most rapidly once the system can begin to be adapted by small changes which preserve most of the already hard-won useful characteristics of the model. But before one can obtain a model which has arrived at this critical condition, a researcher or his predecessors may have had to go through many cycles of developing a complete model, discarding it, and starting all over again.

8.2 Small Models and Large Models

Small models and large models each have their strengths and weaknesses. A small model which has only a small number of variables and relations may be easier to understand, but it may not have a sufficiently rich behavior that its behavior can be compared meaningfully with the system it is to emulate. Too many assumptions may be required to obtain a small model. Too many different causes and effects may be bound up in too few equations and variables. Effects not explicitly represented may perturb the estimation of the relations in the system and obscure their interpretation. Personal experience with the microscopic behavior of socioeconomic systems may not offer much insight to help trace the errant behavior to its source in the structure and/or estimation of the model.

A large model may display richer behavior and explicitly consider many variables and effects which may be obscured in a small model. However, a large model can have so many degrees of freedom that it becomes very difficult to do the necessary estimation, diagnosis, and refinement. The simulation of a large model requires the use of computers and care in the analysis of data handling and solution techniques. The availability of canned simulation programs has done much to make the simulation of large models more practical [Holt et al.: 1964].

Partitioning and tearing can be used to break large models into smaller ones to facilitate the simulation, verification, diagnosis, and refinement of large models.

The techniques required to reduce a large model into smaller ones parallels the techniques used to formulate models, for what we are actually doing is formulating a smaller model of one aspect of the larger model. Therefore, we shall consider several pertinent features of the more general problem of formulating models.

8.3 Isolation

Usually we are concerned about modeling only certain aspects of a complex system. To model realistically all aspects of a complex system at the same time may be too ambitious. Therefore, we are concerned with the problem of how to isolate that part of the system we wish to model from the rest of the system. We must furnish ways to provide for the effects of the rest of the system on the part we are modeling.

A common method of isolation is the use of statistics. Statistical techniques analyze data taken over many observations of the system in order to estimate the free parameters of the model, resolving the variance in the data into a component explained by the model and a random component which cannot be explained by the model.

Another approach to isolation is the use of exogenous variables. These variables represent the effect of the rest of the system on the subsystem to be studied. There are several procedures that may be used to represent such exogenous variables during the simulation:

1. Sometimes the exogenous variables represent policy variables which may be controlled by someone. Then the modeler can simulate his model, making various assumptions about how these policy variables are applied.
2. Often one can represent the exogenous variables by using another model. It may be possible to choose the subsystem to be modeled so the behavior of the model is not sensitive to errors which occur in the modeling of the exogenous variables. Sometimes one can use a time series or just an average value or constant to replace an exogenous variable.
3. When diagnosing a model, one can make a simulation over a period of historical time using values for the exogenous variables which had been recorded in the real system.

Procedures 2 and 3 are pertinent to our use of tearing to derive small models from large models.

8.4 Recursive Models and Simultaneous Models

Another consideration in the formulation of models is whether the model is recursive or simultaneous. A recursive model is one without circuits, while a simultaneous model contains circuits. If when a simultaneous model is partitioned more than one block remains, the model is called *block recursive* or *block triangular*.

Generally, the rules of causality should insure that if the state function were properly chosen the real world would look recursive, i.e., causality contains no circuits. However, circuits can be introduced into the model as a consequence of how the system is observed. For example, there may not be sufficient data or computing power to avoid making some aggregation over time. This aggregation can introduce the appearance of simultaneity in the model. If a effects b and b in turn affects a at a later point in time as follows;

$$a_t \underset{dt}{\gg} b_{t+1} \underset{dt}{\gg} a_{t+2}$$

then the model will be recursive if the time steps it sees are small compared to dt, but will appear simultaneous if the time steps it sees are large compared to $2 * dt$. If we try to save computation by using longer time steps, we may introduce complications which cost us computing time because of simultaneity.

Simon (1957) has used a technique equivalent to our partitioning for the study of causality and simultaneity. We represent "*A* causes *B*" as "*A* precedes *B*." If *A* and *B* are in a common block, then *A* and *B* are simultaneous. Wold (1960) proposes that econometric models should be formulated using sufficiently short time steps and such state descriptions that the model is recursive with recognizable causal chains.

Recursive models are generally easier than simultaneous models to diagnose and refine. One can diagnose each relation in order, knowing that it depends only upon variables produced by relations which have already been diagnosed and refined. The situation is more complicated when a number of relations have to be considered simultaneously. This can result in the "sick horse problem"; that is, the model can tell us it is sick, but it will not say where or why it hurts. Within a block an error in one equation will have an effect on all variables in that same block. Thus it is difficult to trace the causes from the symptoms.

Partitioning is used to obtain a block recursive system where each block represents a simultaneous submodel, with the submodels arranged recursively. Later in this chapter we discuss how tearing can be used to introduce exogenous variables in the blocks of simultaneous models to make them look recursive.

When modeling systems in engineering, the model is usually developed from first principles such that there is no lower limit on the time step inherent in the formulation of the model. Thus models can be formulated to retain this recursive characteristic and avoid the problems due to simultaneity. However, dealing with social systems one must estimate the parameters of the model using statistical procedures. The modeler may not have available data sufficiently replicated and detailed in time to be able to estimate a model on such a time scale as to retain its recursive character. Much economic data exists only on a quarterly or annual basis. But the process that the econometrician is attempting to model may transpire in a period much shorter than that. This means he may not be able to obtain a recursive model and thus does not have the consequent luxury of estimating, verifying, diagnosing, and refining his relations one at a time.

8.5 Partitioning Models

Our output assignment algorithm presented in Figure 5.9 and the partitioning algorithm of Procedure 3.2 and Figure 3.4 have been incorporated into computer programs for the simulation and study of econometric models represented by systems of difference equations.* Partitioning is used to obtain a block recursive ordering. Then only the equations within each block need be solved simultaneously, and the sets of equations within each block can be solved one block at a time. For example, in the simulation of the Brookings-SSRC quarterly model of the United States economy [Duesenberry et al.:65] the model containing 359 simultaneous equations in as many endogenous variables was partitioned such that the largest set of equations to be solved simultaneously was 181.

8.6 Tearing Models

Let us consider the model in Figure 8.1a. It consists of 5 equations in 5 current endogenous variables. Part (b) shows the structural matrix for the equations of the model with outputs assigned; in part (c) the equations and

f_1 (a,b)   = 0
f_2 (d,e)   = 0
f_3 (c)   = 0
f_4 (a,b,c,d) = 0
f_5 (a,c,d,e) = 0
      (a)

(b)

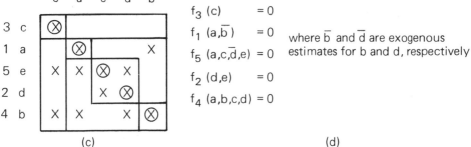

(c)

f_3 (c)   = 0
f_1 (a,$\bar{b}$)   = 0
f_5 (a,c,$\bar{d}$,e) = 0   where $\bar{b}$ and $\bar{d}$ are exogenous estimates for b and d, respectively
f_2 (d,e)   = 0
f_4 (a,b,c,d) = 0

(d)

Figure 8.1

*The Simulate A program developed at the Social Systems Research Institute of the University of Wisconsin [Holt et al.:64] and the TROLL program developed at M.I.T. [Eisner:69].

variables have been reordered in the structural matrix as a consequence of partitioning and tearing. There is a block representing a submodel of one equation followed by a block representing a submodel of four equations. This latter submodel can be broken into two submodels of one equation each and a submodel of two equations if we were to remove the occurrence of variable b from equation 1. The variable b in equation 1 caused equation 1 to be preceded by equation 4, which is solved for variable b, thus causing a circuit. This circuit can be broken if we were to use an exogenous value for variable b as it appears in equation 1 rather than using the value computed in equation 4 (see part [d]). As discussed above, this could be done by using actual values for variable b recorded during some historical period of time while we simulate the behavior of the model during that same historical period. Let us assume that we also use exogenous values for variable d in equation 5. We can study, diagnose, and refine the behavior of each equation one at a time until the model generates the correct values for b and d. Then we may restore the endogenous calculation of these variables and might expect that the model will behave properly.

8.7 Revised Klein-Goldberger Model of the U.S. Economy

The first application we made of the analysis of a system of equations by partitioning and tearing was to the Klein-Goldberger (1955) revised model of the United States economy [Steward:63]. A revised tearing due to suggestions made by Professor Goldberger appears as an example in section 2.7 and the accompanying figures 2.8–2.11 and in section 3.3. This example is used again in the Appendix, figures A.1 and A.2, to illustrate the input and output of the TERABL computer program.

Figure 8.2 lists the definitions of the endogenous and exogenous variables. (It will be remembered from section 3.3 that the equations are solved for the current values of the endogenous variables, using as givens the lagged values of the endogenous variables which have been solved for in earlier time steps and exogenous variables which are provided by some external process such as government policies or experimenters' assumptions.) The equations are listed in Figure 8.3. The small letters above the current variables in the equations mean: o—output endogenous, e—non-output endogenous, x—exogenous. In our analysis we have ignored all but the current endogenous variables so that we are looking at dependencies which occur on the time scale used for collecting the data used in the estimation of the model. These equations are the basis for the program input in Figure A.1. The numbers above the occurrence of torn variables in the equations indicate the level at which they are torn.

Figure 8.2
Klein-Goldberger Model
The endogenous variables of the Klein-Goldberger Model are:

p = price index of gross national product (1939 base: 100)
C = consumer expenditures in 1939 dollars
W_1 = deflated private employee compensation
p = deflated nonwage nonfarm income
S_p = deflated corporate savings
A_1 = deflated farm income, excluding government payments
L_1 = deflated end-of-year liquid assets held by persons
I = gross private domestic capital formation in 1939 dollars
D = capital consumption charged in 1939 dollars
i_L = average yield on corporate bonds
K = end-of-year stock of private capital in 1939 dollars
L_2 = deflated end-of-year liquid assets held by enterprises
P_C = deflated corporate profits
B = deflated end-of-year corporate surplus
Y = deflated national income
N_W = number of wage and salary earners
w = index of hourly wages (1939 base: 122.1)
F_1 = imports of goods and services in 1939 dollars
p_A = index of agricultural prices (1939 base: 100)
i_s = average yield on short-term commercial paper
W_2 = deflated government employee compensation
T_W = deflated personal and payroll taxes less transfers associated with wage and salary income
T_p = deflated personal and corporate taxes less transfers associated with nonwage nonfarm income
T_A = deflated taxes less tranfers associated with farm income
N_p = number of persons in the United States
T_C = deflated corporate income taxes
t = time trend in years
h = index of hours worked per person per year (1939 base: 1.00)
N_G = number of government employees
N_E = number of nonfarm entrepreneurs
N_F = number of farm operators
N = number of persons in the labor force
p_I = index of prices of imports (1939 base: 100)
F_A = index of agricultural exports (1939 base: 100)
R = excess reserves of banks as a percentage of total reserves
G = government expenditures for goods and services in 1939 dollars
F_E = exports of goods and services in 1939 dollars
T = deflated indirect taxes less subsidies
A_2 = deflated farm income, from government payments

Figure 8.3

The equations for the Klein-Goldberger Model are:

(1) $\overset{o}{C}_t = -22.26 + 0.55(W_1 + \overset{x}{W_2} - \overset{x}{T_W})_t + 0.41(\overset{e}{P} - \overset{x}{T_P} - \overset{e}{S_P})_t + 0.34(\overset{e}{A_1} + \overset{x}{A_2} - \overset{x}{T_A})_t + 0.26C_{t-1} + 0.072(L_1)_{t-1} + 0.26(\overset{x}{N_P})_t$

(2) $\overset{o}{I}_t = -16.71 + 0.78(P - T_P + A_1 + A_2 - T_A + D)_{t-1} - 0.073K_{t-1} + 0.14(L_2)_{t-1}$

(3) $(\overset{o}{S_P})_t = -3.53 + 0.72(\overset{e}{P_C} - \overset{x}{T_C})_t + 0.076(P_C - T_C - S_P)_{t-1} - 0.028B_{t-1}$

(4) $(\overset{o}{P_C})_t = -7.60 + 0.68\overset{e}{P_t}$

(5) $\overset{o}{D}_t = 7.25 + 0.10\,\overset{8}{\dfrac{K_t + K_{t-1}}{2}} + 0.044(\overset{9}{Y} + \overset{x}{T} + \overset{o}{D} - \overset{x}{W_2})_t$

(6) $(\overset{o}{W_1})_t = -1.40 + 0.24(\overset{9}{Y} + \overset{x}{T} + \overset{e}{D} - \overset{x}{W_2})_t + 0.24(Y + T + D - W_2)_{t-1} + 0.29t$

(7) $(\overset{9}{Y} + \overset{x}{T} + \overset{e}{D} - \overset{x}{W_2})_t = -26.08 + 2.17[h(\overset{x}{N_W} - \overset{o}{N_G}) + \overset{x}{N_E} + \overset{x}{N_F}]_t + 0.16\,\overset{e}{\dfrac{K_t + K_{t-1}}{2}} + 2.05t$

(8) $\overset{x}{w_t} - w_{t-1} = 4.11 - 0.74(\overset{x}{N} - \overset{e}{N_W} - \overset{x}{N_E} - \overset{x}{N_F})_t + 0.52(p_{t-1} - p_{t-2}) + 0.54t$

(9) $(\overset{o}{F_1})_t = 0.32 + 0.0060(W_1 + \overset{x}{W_2} - \overset{x}{T_W} + \overset{e}{P} - \overset{x}{T_P} + \overset{e}{A_1} + \overset{x}{A_2} - \overset{x}{T_A})_t + \dfrac{\overset{e}{P_t}}{(\overset{e}{p_t})_t} + 0.81(F_1)_{t-1}$

$$(10)\ \overset{o}{(A_1)_t}\, \frac{\overset{e}{P_t}}{(\overset{e}{p_A})t} = -0.36 + 0.054(\overset{e}{W_1} + W_2 - \overset{x}{T_w} + \overset{e}{P} - \overset{x}{T_p} + \overset{e}{P} - \overset{x}{T_p} - \overset{e}{S_p})\frac{\overset{e}{P_t}}{(\overset{e}{p_A})_t}$$
$$-0.007(\overset{e}{W_1} + W_2 - \overset{x}{T_w} + \overset{e}{P} - \overset{x}{T_p} - \overset{e}{S_p})_{t-1}\frac{P_{t-1}}{(\overset{e}{p_A})\,t_{-1}} + 0.012(\overset{x}{F_A})_t$$

$$(11)\ \overset{o}{(p_A)_t} = -131.17 + 2.32\overset{e}{p_t}$$

$$(12)\ \overset{o}{(L_1)_t} = 0.14(\overset{e}{W_1} + \overset{x}{W_2} - \overset{x}{T_w} + \overset{e}{P} - \overset{x}{T_p} - \overset{e}{S_p} + \overset{e}{A_1} + \overset{x}{A_2} - \overset{x}{T_A}) + 76.03(\overset{e}{i_L} - 2.0)_t$$

$$(13)\ \overset{o}{(L_2)_t} = -0.34 + 0.26(\overset{e}{W_1})_t - 1.02(\overset{e}{i_l})_t - 0.26(\overset{e}{p_t} - p_{t-1}) + 0.61(L_2)_{t-1}$$

$$(14)\ \overset{o}{i_l} = 2.58 + 0.44(i_s)_{t-3} + 0.26(i_s)_{t-5}$$

$$(15)\ 100\,\frac{\overset{o}{(i_s)_t} - (i_s)_{t-1}}{(i_s)_{t-1}} = 11.17 - 0.67R_t$$

$$(16)\ \overset{e}{C_t} + \overset{e}{I_t} + \overset{x}{G_t} + (\overset{e}{F_E})_t - (\overset{e}{F_I})_t = \overset{o}{Y_t} + \overset{x}{T_t} + \overset{e}{D_t}$$

$$(17)\ \overset{e}{(W_1)_t} + \overset{x}{(W_2)_t} + \overset{o}{P_t} + (\overset{8}{A_1})_t + (\overset{9}{A_2})_t = Y_t$$

$$(18)\ \overset{x}{\underline{h}}\,\overset{e}{w_t}\overset{}{(N_w)_t} = \overset{e}{(W_1)_t} + \overset{x}{(W_2)_t}$$
$$\overset{o}{P_t}$$

$$(19)\ \overset{o}{K_t} - K_{t-1} = \overset{e}{I_t} - \overset{e}{D_t}$$

$$(20)\ \overset{o}{B_t} - B_{t-1} = (\overset{e}{S_p})_t$$

It may also be instructive to add marks to the matrix representing the occurrence of the lagged values of the endogenous variables to show what affects what in the equilibrium which occurs over a longer period of time.

In the input and output of the program (figures A.1 and A.2) the variable identifications appear with capital letters in the first column with lower-case letters and subscripts shifted to the right. The equations and variables are also assigned numbers.

When working with systems of equations, if there are no marks on the diagonal, the procedure of Figure 5.9 is used to select an output assignment and permute the equations to bring the assigned outputs to the diagonal. As shown in section 5.2 the partitioning is unique and independent of any arbitrariness of this assignment. However, as discussed in section 5.4 the selection of the output assignment does affect the process of selecting tears. Thus, when the modelers have written the equations with a distinct, current endogenous dependent variable to the left of the equal sign, we have assigned this variable as the output of that equation. We assumed that the equations were so written that this assignment corresponds to a causal interpretation where changes in the independent variables cause changes in the dependent variable. Where the modelers did not so indicate a distinct dependent variable, we made an arbitrary output assignment. The TERABL program was forced to accept this assignment by so numbering the equations and variables that the desired outputs already occur on the diagonal. Then the program accepts this as its output assignment.

Parts (b) and (c) of Figure A.2 show the results of partitioning the model. There remains a block of 14 equations. Let us first look for tears in this block that appear good from a strictly structural point of view. Then we will evaluate the various possible tears from the point of view of their economic meaning.

Consider the shunt diagram summary in Figure A.2f. Under the NS column we find that each arc in the principal circuit has at least 8 shunts. Arcs (16,1) and (5,16) have the least number of shunts with 8 and 9, respectively. We do not find an arc in the principal circuit with a particularly high FS. A high value of FS would have indicated that all the shunts have a high index and thus tearing such an arc in the principal circuit would be expected to leave small remaining blocks. A particularly high value of NS would indicate that reassigning the outputs in order to reverse the principal circuit might yield a good setup for tearing. The highest value of NS is 17, indicating that after reversing the principal circuit we may expect an arc with only 6 shunts (the total number of shunts minus the largest value of NS). This change in minimum number of shunts would not be very significant, and furthermore, we are reluctant to reverse the principal circuit since we feel that the outputs are now assigned in a reasonable way. Under NS-NB we note that arc (5,16) has only 3 shunts which would not be broken by tearing arcs exiting 16. This would correspond to tearing marks in column 16. And under NS-NE we note that arc (16,1) has only 3 shunts which would not be broken by tearing marks entering 16. This would correspond to tearing marks in row 16. Variable 16, National Income, then appears to be a key variable. We could expect to obtain a good reduction in the size of the blocks which remain if we were to tear either the arcs entering or the arcs exiting variable 16.

Now let us look to see which variables might make economic sense as feedback variables, with particular attention focused on variable 16, National Income. It appears that National Income would make a good feedback variable because it sums the various sources of income, which is then distributed back into the economy. It would appear to make sense to introduce such a variable as exogenous and then see how the model causes that money to be redistributed.

In the shunt diagram in Figure A.2d we have drawn a line representing the tearing of the arc (5,16) in the principal circuit. The line intersects the shunts parallel to this arc. We choose to tear the first arc in each parallel shunt which has a B opposite the 16. These arcs each exit 16 and thus correspond to marks in column 16 in the precedence matrix. Tearing the marks exiting 16 has the effect of moving variable 16 toward the end of the block. Tearing arcs entering 16, i.e., tearing the last arcs of the shunts with E's opposite 16, corresponds to tearing marks in row 16 and would have the effect of moving variable 16 toward the beginning of the block. Since we interpret National Income as a summary variable, we prefer tearing the exiting arcs and moving it toward the end of the block. Furthermore, tearing marks in just this one column would mean that we can use the same exogenous variable in each equation where one of these marks is torn.

A check mark has been placed above each of these shunts which are torn; x's are placed above the parallel shunts which have not been torn. Note that the arc (6,16) appears in four of these parallel shunts. Thus, tearing this one arc has the effect of breaking all four shunts.

We remember that our necessary criterion for all circuits to be broken is that there remains no one untorn shunt which is parallel to all torn arcs in the principal circuit. At this point we have 3 remaining shunts parallel to the one torn arc in the principal circuit. We might proceed then to tear another arc in the principal circuit, or to tear the shunts themselves, or some combination. We choose instead to defer this decision, to make the tears discussed above, then partition and form a new shunt diagram, before making any conclusion about further tears. This is always a prerogative which we have. Note that if we consider the two untorn parallel shunts of index 10, no arc in the principal circuit is parallel to gaps in both shunts.

Thus, we can already see that more than one additional tear will be required to eliminate all circuits.

After we set these tear marks to level 9 and partition, we find that we are left with a block of size 2 and a block of size 4; see Figure A.2h–i. In the block of size 2 we must choose to tear one of two marks. Not seeing any advantage one way or the other, we made the choice arbitrarily.

Looking at the shunt diagram or shunt diagram summary in parts (j) and (k), we can see that tearing the arc (17,10) in the principal circuit would leave no shunt parallel to this torn arc. However, in discussing this with Professor Goldberger, he preferred to tear (10,17) and (10,3) so that Other Income (17) and Corporate Savings (3) feed back to Farm Income (10) on the grounds that these tears represented the longer time constant dependencies. (Note that we can go back and see that these same tears could have been recognized from the first shunt diagram and shunt diagram summary in parts (d) and (e) if we had chosen to make the decision at that time.)

Making these additional tears at level 8 and partitioning produces the ordered precedence matrix in Figure A.2n. Using this ordering we might introduce exogenous variables for National Income, Corporate Savings, Other Income, and Private Capital where these variables appear above the diagonal.

8.8 Input-Output Models

Input-Output models are represented by a matrix or network which shows the flow of the value of goods from one industrial sector to another. Unlike simultaneous equation models, input-output models do not require the assignment of outputs.

Let x_j be the value of the output of goods from industrial sector j, c_i be the value of goods from indistrial sector i which are consumed without contributing to another sector in the system, and a i_j be the value of goods from sector i used to produce a unit value of goods by sector j. Then

$$\sum_{j=1}^{j=n} a_{ij} * x_j + c_i$$

is the value of goods from industrial sector i required to supply the set of outputs x_j from sectors $j = 1, \ldots ,n$, plus the consumption c_i. The case of static equilibrium can be represented by the matrix equation:

$$x = Ax + c \qquad \text{where}$$
$$x = \text{the vector } [x_j]$$
$$c = \text{the vector } [c_i]$$
$$A = \text{the matrix } [a_{ij}]$$

Dorfman, Samuelson and Solow [53] call A decomposable if it can be partitioned such that more than one block occurs on the diagonal. Each block represents a subeconomy with a two-way sequence of flows of the value of goods between any two industries within the same block, and no flow from any sector in one block to a sector in an earlier block.

We propose that we can further order the industries within a block by assigning discrete level numbers to the dollar flows to obtain a precedence matrix which we can then partition by levels as in Procedure 3.4. The higher the dollar value of the flow the lower the level of the corresponding mark in the precedence matrix. Then the smaller dollar flows are torn before the larger dollar flows. Note that we are partitioning based upon circuits resulting from marks at a given level or below. The circuits imply that the dollar value of the flow of goods from each industrial sector to the next in the circuit is at least the minimum flow required to admit that flow to this level.

8.9 Leontief's Input-Output Model

We have used Leontief's input-output model as it appeared in *Scientific American* (1965) to illustrate the application of partitioning and tearing analysis to input-output models.

Leontief states that "the 81 industrial sectors have been 'triangulated', that is, they have been arranged in accordance with the hierarchy of interindustrial dependence." He does not state any more precisely how he has made this choice of ordering. We assume that he used a combination of looking at the numbers, exercising some judgment, and a great deal of playing with many rearrangements.

We randomized the initial ordering of the industrial sectors, then applied the techniques of partitioning by levels as in Procedure 3.4 to order the sectors. We set up seven discrete levels and assigned the dollar values of the transactions to these levels according to Table 8.1. The higher the dollar transaction, the lower the level number assigned; so that when partitioning by levels, the smaller dollar values are removed earlier and the higher dollar values are removed later.

INPUT-OUTPUT MODEL   BEFORE PARTITIONING
    PAGE 1/ 1

#	Sector	Row
63	CRUDE PETROLEUM,NATURAL GAS	1
41	STONE & CLAY MINING & QUARRY	2
32	MACHINE SHOP PRODUCTS	3
74	COMMUNICATION-NOT RADIO,TV	4
66	STATE,LOCAL GOVT ENTERPRISES	5
49	MISC TEXTILE GOODS,FLOOR COVE	6
44	PAPERBOARD CONTAINERS & BOXES	7
19	SCIENTIFIC, CONTROL INSTRUMENT	8
5	APPAREL	9
58	CHEMICALS, SELECTED CHEM PROD	10
12	MISC TRANSPORTATION EQUIPMENT	11
27	GEN INDUSTRIAL MACH, EQUIP	12
50	RUBBER,MISC PLASTICS PRODUCTS	13
18	HOUSEHOLD APPLIANCES	14
45	PAPER,ALLIED PROD-NOT CONTAIN	15
2	MISC FURNATURE AND FIXTURES	16
23	CONSTRUCT,MINE,OIL FIELD EQUI	17
47	LUMBER,WOOD PRODS-NOT CONTAIN	18
43	GLASS & GLASS PRODUCTS	19
51	FABRICS,YARN,THREAD MILLS	20
22	ENGINES AND TURBINES	21
17	SERVICE INDUSTRY MACHINES	22
29	ELECTRIC INDUSTRIAL EQUIP	23
76	REAL ESTATE AND RENTAL	24
24	MISC ELEC MACH,EQUIP,SUPPLIES	25
68	AUTO REPAIR AND SERVICE	26
16	OPTICAL, OPTH, PHOTO EQUIP	27
42	PRINTING & PUBLISHING	28
21	FARM MACHINERY AND EQUIPMENT	29
15	MISC MANUFACTURING	30
14	MATERIALS HANDLING MACH&EQIP	31
54	LIVESTOCK AND PRODUCTS	32
48	FORESTRY,FISHERY PRODS	33
36	PRI NONFERROUS METAL MANUFACT	34
79	OFFICE SUPPLIES	35
70	AMUSEMENTS	36
53	LEATHER TANNING INDUS PRODS	37
65	TRANSPORTATION,WAREHOUSING	38
78	RESEARCH AND DEVELOPMENT	39
67	MOTELS,PERSONAL,REPAIR SERVIC	40
69	RADIO,TV BROADCASTING	41
75	BUSNESS SERVICE	42
6	MISC FABRICATED TEXTILE PRODS	43
13	RADIO,TV,COMMUNICATION EQUIP	44
71	MED,EDUCATION SERVICE:NONPROF	45
61	ELECTRICITY,GAS,WATER	46
3	HOUSEHOLD FURNATURE	47
39	IRON & FERROALLOY ORES MINING	48
20	OFFICE,COMPUTING,ACCOUNT MACH	49
59	CHEM,FERTILIZER,MINERAL MINE	50
1	FOOTWARE AND LEATHER PRODUCTS	51
26	MOTOR VEHICLES AND EQUIP	52
7	DRUGS, CLEANING, TOILET PREPS	53
25	METAL WORK MACH AND EQUIP	54
64	FEDERAL GOVT ENTERPRISES	55
37	NONFERROUS METAL ORES MINING	56
80	BUSINESS TRAVEL,ENTERTAIN,GIF	57
4	TOBACCO MANUFACTURES	58
72	WHOLESALE,RETAIL TRADE	59
9	SPECIAL INDUSTRY MACH& EQUIP	60
52	PAINTS,ALLIED PRODS	61
33	METAL CONTAINERS	62
62	COAL MINING	63
73	FINANCE, INSURANCE	64
57	PLASTIC,SYNTHETIC MATERIALS	65
31	HEAT,PLUMB,STRUCT METAL PRODS	66
46	WOODEN CONTAINERS	67
60	PETROLEUM REFIN,RELATED INDUS	68
56	AGRIC,FORESTRY,FISHERY SERVIC	69
81	SCRAP,USED,SECONDHAND GOODS	70
38	PRI IRON & STEEL MANUFACTURE	71
40	STONE AND CLAY PRODUCTS	72
34	STAMPING,SCREW MACH PROD,BOLT	73
55	MISC AGRICULTURAL PRODUCTS	74
10	ORDANCE AND ACCESSORIES	75
8	FOOD AND KINDRED PRODUCTS	76
30	ELECTRONIC COMPONENTS & ACCES	77
35	OTHER FABRICATED METAL PRODS	78
77	MAIN,REPAIR CONSTRUCTION	79
28	ELECTRIC LIGHTING,WIRING EQUI	80
11	AIRCRAFT AND PARTS	81

Figure 8.4a
TERABL Output—Leontief's Input-Output Model

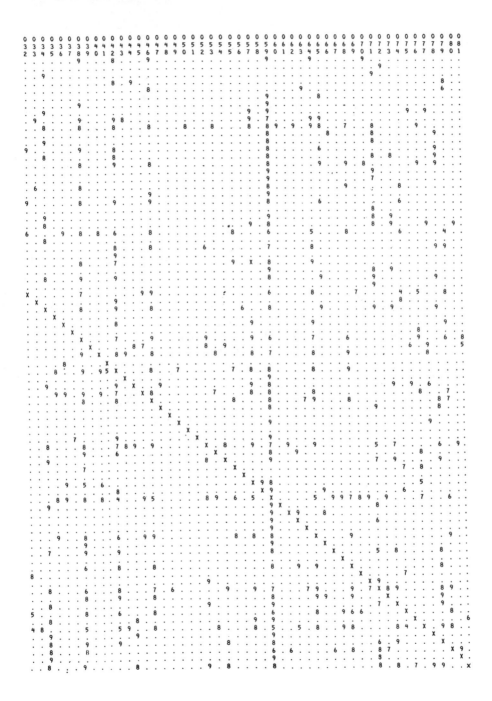

Dollars (millions)	Level
100 to 199	9
200 to 599	8
600 to 999	7
1000 to 1999	6
2000 to 4999	5
5000 up	4

TABLE 8.1:   Assignment of Levels to Dollar Flows for Leontief's Input-Output Model

Figure 8.4 shows the matrix ordered by this procedure. The number on the left shows the number in Leontief's ordering with the smaller numbered sectors tending to supply the higher numbered sectors. The number on the right shows the numbers as they were ordered randomly to begin the ordering by our process. Because of the use of discrete levels, variations of the dollar values within a level could not be expected to influence the ordering. Thus there will be some effect due to the original ordering on the order of sectors within a block. However, which sectors occur in which blocks would not be affected.

It will be noted that our ordering varies quite a bit from Leontief's. Note that several sectors brought together in our ordering make sense as interrelated economic sectors, but were not so related in Leontief's ordering, such as (76) Real Estate and Rental with (73) Finance, Insurance; or (55) Misc Agricultural Products, (54) Livestock and Products, and (8) Food and Kindred Products.

```
INPUT-OUTPUT MODEL
ORDER   81
HIGHEST LEVEL MARK TO BE CONSIDERED   9
LARGEST BLOCK FOR WHICH SHUNT DIAGRAM IS NOT PRINTED   2
LARGEST INDEX SHUNT TO BE CONSIDERED   81
```

```
LEVEL   9
        1   1 74   1   1   1   1   1
```

```
LEVEL   8
        1   1   1   1   1   1   1   1   1   1   1 52   1   1   1   1   1   1   1   1
        1   1   1   1   1   1   1   1   1   1
```

```
LEVEL   7
        1   1   1   1   1   1   1   1   1   1   1   1   1   1   1   1   1   1   1   1
        1   1   1   1   1   1   1   1 28   1   1   1   1   1   2   1   1   1   1   1
        1   1   1   1   1   1   1   1   1   1   1   1   1
```

```
LEVEL   6
        1   1   1   1   1   1   1   1   1   1   1   1   1   1   1   1   1   1   1   1
        1   1   1   1   1   1   1   1   1   1   1   1   1   1 21   1   1   1   1   1
        1   1   1   1   1   1   1   1   1   1   1   1   1   1   1   1   1   1   1   1
        1
```

```
LEVEL   5
        1   1   1   1   1   1   1   1   1   1   1   1   1   1   1   1   1   1   1   1
        1   1   1   1   1   1   1   1   1   1   1   1   1   1   1   1   1   1   1   1
        1   1   1   1   1   1   2   7   1   1   1   1   1   1   1   1   1   1   1   1
        1   1   1   1   1   1   1   1   1   1   1   1   1   1
```

```
LEVEL   4
        1   1   1   1   1   1   1   1   1   1   1   1   1   1   1   1   1   1   1   1
        1   1   1   1   1   1   1   1   1   1   1   1   1   1   1   1   1   1   1   1
        1   1   1   1   1   1   1   1   1   1   1   1   1   1   1   1   1   1   1   1
        1   1   1   1   1   1   1   1   1   1   1   1   1   1   1   1   1   1   1   1
        1
```

```
LEVEL   0
        1   1   1   1   1   1   1   1   1   1   1   1   1   1   1   1   1   1   1   1
        1   1   1   1   1   1   1   1   1   1   1   1   1   1   1   1   1   1   1   1
        1   1   1   1   1   1   1   1   1   1   1   1   1   1   1   1   1   1   1   1
        1   1   1   1   1   1   1   1   1   1   1   1   1   1   1   1   1   1   1   1
        1
```

Figure 8.4b

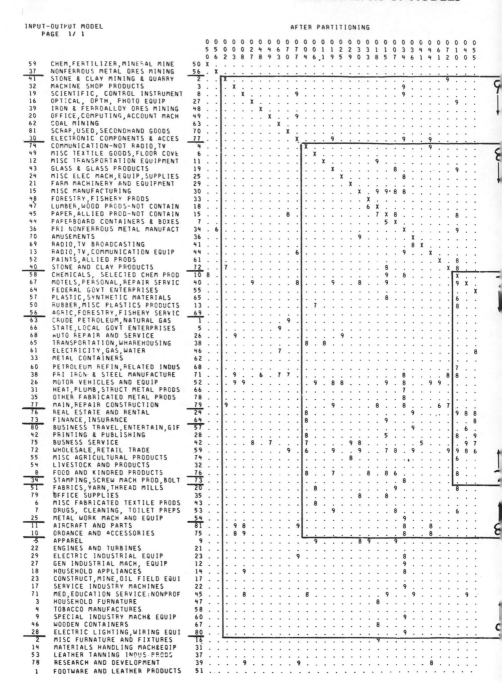

Figure 8.4c

AFTER PARTITIONING

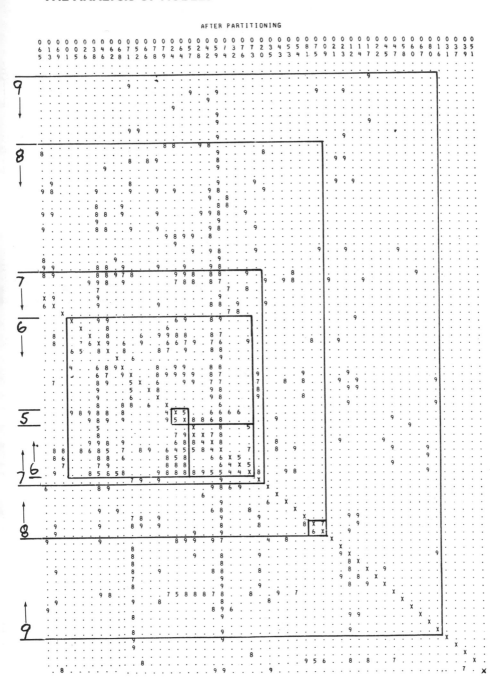

9
Government, Complexity and the Elephant Problem

9.1 Government and Complexity

In government, proposed actions have to be justified to many people. Assume you are a legislator; your proposal must be justified to other legislators if the legislation is to be passed, to the electorate if you are to be re-elected, to campaign contributors and lobbyists if you are to afford the next election, and to administrators who must implement the legislation. It is nearly impossible to preserve and defend an understanding of subtleties through all these layers of interaction. As a consequence, government often acts by considering only the most immediate, direct, and obvious consequences of its actions.

The behavior of our world involves more and more interaction. As the population increases, as the rates of use of energy and resources increase and the finite limits of these resources become more obvious, it is no longer possible to do anything without affecting a great number of people and producing many subtle consequences. This greater interaction makes our lives more complex.

As one example of many, consider the siting of energy plants, e.g., coal or nuclear electrical generating stations, oil refineries, liquefied natural gas ports, etc. In each 15-year period the rate of energy use doubles; thus, during this period we must build as many new energy plants as had existed before just to handle the growth. With an increasing population and finite space, it is now impossible to build plants that will not impact through pollution or risk on a large number of people. Even more people are affected by the cost of reducing this pollution or risk. There can result a loss of jobs in one area, and the creation of new jobs with different training requirements in another. Dependence on foreign fuel supplies makes the effects international. A few years ago a few energy plants could be built in isolation with none of these ramifications.

Past growth of population density in cities is now yielding horrendous social and financial problems. We are seeing severe and complex economic and employment problems as the number of jobs grows but the number seeking jobs grows faster. We face the problem of inflation. More and more people must share less space and shrinking resources. Pollution has become a serious problem. Solving one problem can make another one worse.

As the various facets of our lives and businesses become more intertwined, the problems we face as individuals, and particularly the problems faced by government, have become more complex. The simple solutions we once could have used, that consider only the most immediate, direct, and obvious effects, no longer work in our more complex world.

Although we are quick to attribute our problems and frustrations to big government, much of the difficulty can be traced to the increased complexity of the problems faced by government. The complexity arises from greater interactions in our society as more and more people share less and less. The greater interdependence means that actions have more indirect effects, which make the consequences more difficult to understand and cope with.

As the problems faced by government have become increasingly complex, the solutions have become less workable and taxpayers more frustrated. Taxpayers have begun denying government the money needed to solve these problems at the very time the problems are becoming more difficult and more necessary to solve.

How do we get a grasp on these complex problems? We need a tool to help us reason through both the direct and indirect consequences of proposed actions. Such a tool is the method of structural modeling as discussed below.

9.2 The Elephant Problem

Let us assume you are a government official responsible for making changes in a complex system to obtain some desired result. Perhaps it's a complex socioeconomic problem such as how to attract investment into the inner city, or how to reverse the migration of school enrollment from the city to the suburbs. You don't understand all aspects of the problem yourself so you bring in consultants who each have an understanding of some important part of the system. After listening to these experts, their opinions and explanations, do you then have a good plan of action that you can defend? Chances are if you have *n* experts you have *n* different recommendations. How do you proceed now? How do you get a hold on putting together the experts' views about the various parts of the system to gain sufficient comprehension of the whole to make a good decision?

Each expert viewing his own particular aspect of the system can be thought of as one of the blind men, each feeling a different part of the elephant. One says it's a tree, others say it's a hose, a wall, a rope, depending upon which part of the elephant they feel. The manager needs a way to put all this together to see an elephant. Only then can he make sense of what he is told about the parts, direct good questions to the experts, and make decisions about what policies to undertake. In short, the manager needs to understand the structure of the system—what affects what—well enough to be able to take command.

9.3 Can We Make a Model?

One might suggest the manager needs a model to integrate the views of the parts into a perspective of the whole. But the usual modeling tools may not be of much use. Many managers are not comfortable with modeling techniques. Others are leary of models because they recognize the limitations and problems with modeling. Managers are justifiably concerned that the experts will use the models to support their own ideas and amaze and confuse the uninitiated. For a manager to make a judgment he must understand the model and its weaknesses. It is a great deal to ask that he become an expert in every field.

A great deal must be known about a system before conventional modeling techniques can be used. Of course, the model can help one learn more about the system. But it has to start with knowing enough to be able to develop a model first.

We facetiously refer to the three levels of modeling sophisitication as follows:

Muddling—This is the lowest level of sophistication and really does not represent modeling at all. It means acting with no clear understanding of the interactions involved in a system and results in significant resource expenditures and poor performance. It is often the best one can do in the absence of better methods, but is not a valid excuse for anyone who has read this chapter.

Meddling—This second level of sophistication is inept modeling which occurs when one thinks he knows how to model a system but is really unaware of the weaknesses and pitfalls of the model. It can also occur when one hopes to get additional funds to do a better job of modeling and thus dare not reveal the problems in the current efforts for fear of losing that money. This condition is fostered by not expecting enough in the long term, but expecting too much in the short term. Muddling can result in poor performance. But meddling can result in that false confidence that leads to commiting sufficient resources to produce a true disaster.

Modeling—This is what we aspire to. The excellence of a model is not to be judged by its size in number of variables, but by its ability to make valid predictions that would otherwise not have been expected, and by how well it leads to a better understanding of the system.

There is an adage that says those who can make models will make models, of whatever they can. But these models may or may not contribute to solving the problem at hand.

We consider below how to develop the structure of the system. The structure shows how the various parts of the system interrelate, which can be important as a means to get started when one intends to do more elaborate modeling. It can also help the manager gain control over how the problem is handled.

9.4 Paradoxical Laws of Systems

Consider the following example [Churchman:68]. A company's executives learned of a technique for solving the transportation problem at a conference and immediately had their analysts model the distribution of goods among their factories, warehouses, and customers. The solution they obtained with a sophisticated analysis using the transportation algorithm showed no significant room for improvement over their existing system. But when someone made a very simple analysis of the larger problem, including the possibility of relocating the warehouses, an opportunity for some very significant improvements was found.

This leads us to the "paradoxical laws of systems":

1. If you consider only a part of the system, some other part you did not consider will very likely hide an effect of critical importance. So you should consider the larger system.
2. When you consider the larger system, the problem becomes more complex.
3. If the problem becomes too complex, you cannot deal with it.

What saves us from this dilemma is that a very simple analysis of the structure of the larger problem will often show opportunities not to be found in even the most sophisticated analysis of only a part of the problem.

9.5 Structural Models

We argue for looking at the structure of the system as a first step before deciding where to focus further attention. The structure consists of a set of conditions and the cause and effect relations between these conditions. The techniques for gathering this data are within the grasp and established skills of good managers. The analysis can be done with the help of a computer program and a technician who has read chapters 2, 3, and 4 of this book. The result of this analysis is just a reorganization of the information which management and the experts originally supplied and understood. But now it appears in a form that helps enhance one's insight into how the pieces of the system act together.

The results of this analysis can be used by management to play what-if games in the cause and effect structure, looking for what we call the jujitsu point in the system. In jujitsu minimum effort is used to manipulate a strong opponent. As we use the term, the *jujitsu point* is where a minimum cost or energy change can alter the behavior of a large complex system in the desired way.

This structural model is a way of showing the relationship of the parts of the system. It is a way of pulling together what one person or a group of people may know about a system to focus and discipline one's thinking while exploring various tactics for solving the problem. It works like algebra—it allows one to express what is known in a formal way and to manipulate it until the consequences lead to a solution or greater insight.

The method discussed here for formulating the problem has been developed by Warfield (1976) and others, who refer to these methods under the name Interpretive Structural Modeling. This chapter discusses our interpretation of their methods of formulating the problem, and uses the techniques developed in this book to extend their methods of analysis.

9.6 An Annotated Example

As an example we present the problem of determining the most cost-effective way for the government to make changes so that business investment will be attracted into the city.*

* The problem and the data used in this example were derived from Malone (1975).

Table 9.1 shows the list of conditions proposed by a set of experts as having an effect on the problem of attracting investment into the inner city. For each condition they have indicated by number the predecessors, i.e., the other conditions which contribute to this condition.

TABLE 9.1: "Difficulty Attracting Investment into City"
Precedence Table

Number	Description of Conditions	Predecessors
1	High crime rate	2, 3, 5, 10, 11, 12
2	Vandalism	3, 5, 6, 7, 10, 11, 12
3	Poor public schools	11,18
4	Traffic congestion	28
5	Poor public services	11, 18, 21
6	Abandoned buildings	14, 18, 19, 22, 26
7	Poor state of property maintenance	2, 9, 14, 18, 19, 22, 26
8	High number of families on welfare	12
9	Lack of incentive to maintain appearance	2, 7, 14
10	Lack of facilities for youth	11, 18
11	Alienation of residents	21, 24, 25
12	Downward income distribution	29
13	Fragmented property ownership	0
14	Excess supply of housing	12
15	Need for low income housing	8
16	High land costs	0
17	High construction costs	0
18	High property taxes	1, 2, 3, 5, 7, 8, 10, 15, 16, 17, 20, 25, 28
19	High insurance costs	6, 9, 20
20	High security requirements	25
21	Delays in getting civic actions	0
22	High borrowing costs	9, 20
23	High land assembly costs	13, 16, 22
24	Uncertainty of civic constraints	0
25	Perception of insecurity	1, 2
26	Inability to finance w/o govt. guarantees	9, 20
27	Lack of adequate parking	28
28	Inadequate mass transit	18, 26
29	Difficulty attracting investment into city	2, 3, 4, 5, 9, 10, 12, 17, 18, 19, 21, 22, 23, 24, 25, 26, 27, 28

Figure 9.1 shows the same information but in matrix form. Each row and its corresponding column represent a condition. Circled *x*'s are used to mark the diagonal. The off-diagonal *x*'s show the relations between conditions. The *x* in column 2 row 25 implies that the column Vandalism

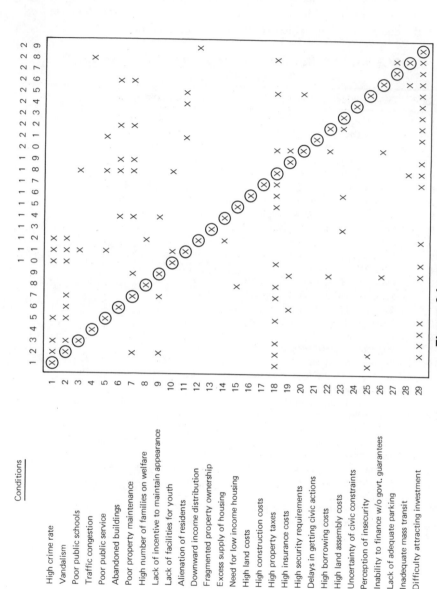

Figure 9.1

Difficulty Attracting Investment: Matrix #1—Original

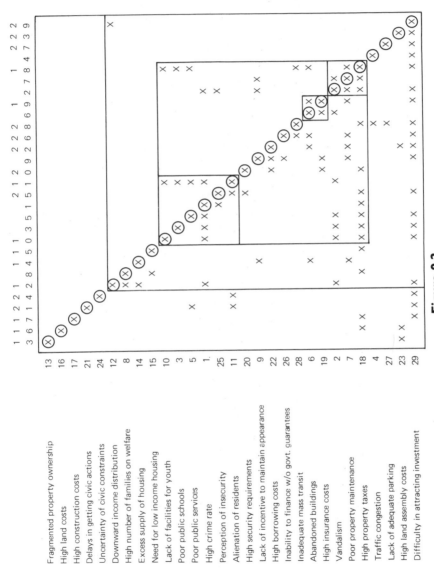

Figure 9.2

Difficulty Attracting Investment: Matrix #2—Matrix #1 Reordered

has an effect on the row Perception of Insecurity. The x's are scattered; it is difficult to see a pattern from this matrix.

Reordering the rows and corresponding columns produces the matrix in Figure 9.2. The techniques to do this reordering are covered in chapters 2, 3, and 4. The reordering helps us make more sense of the information in the original matrix, but we must know how to interpret the matrix.

The boxes on the diagonal show the set of conditions such that each condition in the box affects and is affected by every other condition in the box. To see this, look at the small 2-by-2 box on the diagonal where the rows and columns for conditions numbered 6 and 19 intersect. Condition 6 is Abandoned Buildings, and condition 19 is High Insurance Costs. The x in column 6 row 19 means that 6 (Abandoned Buildings) has an effect on 19 (High Insurance Costs). The x in column 19 row 6 means that 19 (High Insurance Costs) has an effect on 6 (Abandoned Buildings). Thus, within this block each condition affects the other.

Similarly, within the largest block, which involves conditions 12, 8, 14, etc., through 29, every condition in the block has an effect on every other in the block and vice versa. Let us illustrate. We will write (14,9) to represent the mark in column 14 row 9, or that 14 affects 9. Now let us take any two conditions within this block, say 14 and 6. The sequence of marks (14,9), (9,26), and (26,6) show a chain that implies that 14 affects 9 affects 26 affects 6. Similarly, the sequence of marks (6,19), (19,29), (29,12), (12,14) shows there is a path from 6 back to 14. This property is true for every pair of conditions in the block. In fact, this property defines the block.

All the conditions within a block are interrelated. Change any one and all the rest will be affected.

Now let us look at the block within this block, the one that involves the conditions 10, 3, etc., to 18. Look at the mark (29,12), which is outside the smaller block, but still inside the larger block, and above the diagonal. If we were to remove this mark, we would find that we have the block property within the smaller block but no longer within the larger block. So the relation shown by the mark (29,12) is a key to tying together the conditions in the larger block.

The reordering has tended to bring together those conditions which are most closely interrelated. When the experts look at Figure 9.2, which has been reordered, they can see more clearly the structure implied by the information they put into the matrix in Figure 9.1; they are then inclined to reconsider their earlier inputs and make changes. Figure 9.3 shows these changes to Figure 9.2. A y is an element added, a d an element deleted. They may also decide at this time that certain conditions that are closely related and need not be distinguished should be combined as shown by the brackets next to the row numbers. The resulting matrix is shown in Figure

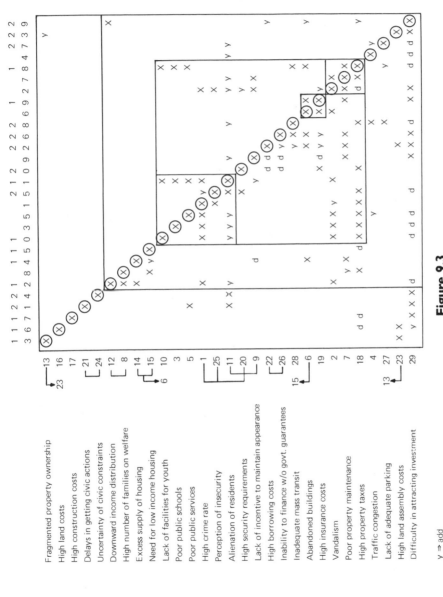

Figure 9.3

Difficulty Attracting Investment: Matrix #3—Changes Made to Matrix #2

y ⇒ add
d ⇒ delete

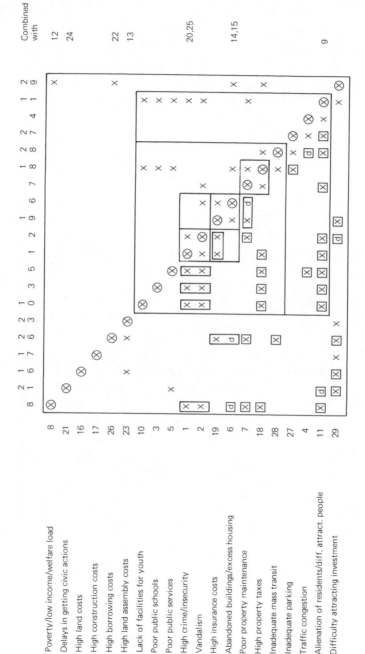

Figure 9.4

Difficulty Attracting Investment: Matrix #4—Matrix #3 Reordered, Redundant Elements Marked

9.4. This matrix is marked up to show some boxes around x's and d's instead of x's, which we will explain below.

Note the mark (28,27) which says that 28 (Inadequate Mass Transit) affects 27 (Lack of Adequate Parking), and the mark (27,4) which says that 27 (Lack of Adequate Parking), affects 4 (Traffic Congestion). Together these imply that 28 (Inadequate Mass Transit), affects 4 (Traffic Congestion). This is an example of the transitive property, i.e., if A affects B and B affects C, then A affects C.

The matrix also contains the element (28,4) which implies that 28 affects 4 directly. This would say that the effect between 28 (Inadequate Mass Transit) and 4 (Traffic Congestion) occurs directly as well as through the intermediate condition Inadequate Parking. It is very easy to put in direct effects when they are not intended. This happens when we fail to note that the effect occurs only through another intermediate condition in our list. If the effect is only through the intermediate condition, it may be possible to intervene by influencing only the intermediate condition. But if the effect also occurs directly, we may have to intervene directly as well as indirectly. Thus we want to be careful not to put in direct effects when we only have indirect effects. So the x in (28,4) has been replaced by a d to show that it will be deleted.

We can have a computer generate all the indirect effects by transitivity. Whenever we find that we have input a mark that is also implied by indirect effects, we stop to ask whether the direct effect occurs in addition to the indirect effect, or whether we failed to recognize the required intermediate condition.

Figure 9.5 is the transitive closure of the matrix in Figure 9.4. By this we mean that all the indirect effects generated by transitivity are shown in the matrix. We have put boxes around elements or sets of elements that were given directly by the input and also implied by transitivity. If on reflection we decided that there was no direct effect, the mark should be deleted. This is shown by a d in figures 9.3 and 9.4. In Figure 9.6 the deletions have occurred.

Figure 9.6 is the result of making these changes to the matrices shown in figures 9.3 through 9.5. We refer to making the changes as massaging the matrix.

Let us see what we can find in Figure 9.6. Our goal is to change condition 29 (Difficulty in Attracting Investment). This condition occurs in the last row and column. We notice marks in column 29, which tell us what other conditions will be affected when we accomplish our goal. We can ignore these marks while considering how to make the changes to accomplish the goal. Let us look for some conditions we might be able to change and see what the consequences may be.

Poverty/low income/welfare load would be a difficult condition to

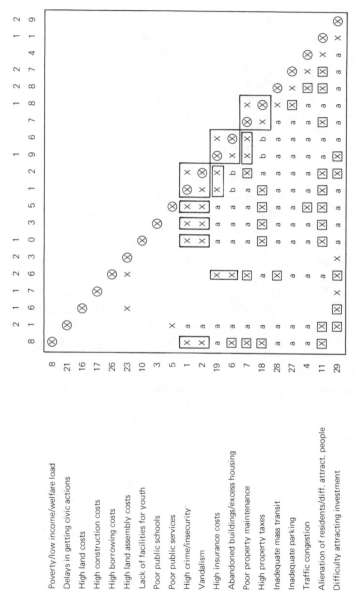

Figure 9.5

Difficulty Attracting Investment: Matrix #5—Transitive Closure to Find Redundant Elements

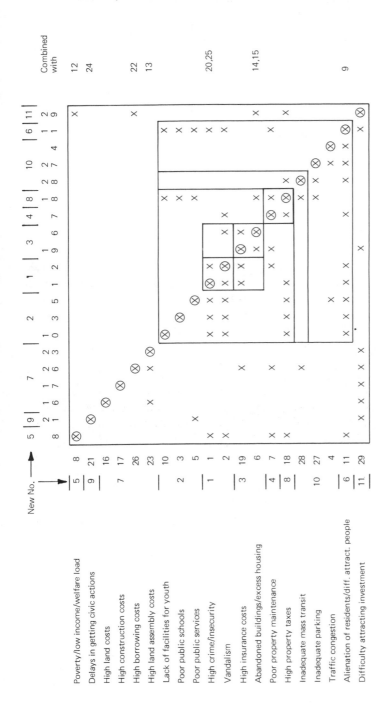

Figure 9.6
Difficulty Attracting Investment: Matrix #6—Matrix #4 (Without Extra Marks)
Showing Groupings to Obtain Reduced Matrix

change directly, but we might hope to change it indirectly through the effects of other conditions. We do see by (29,8) that when we succeed in our goal of attracting investment that poverty, etc. will be affected too.

Delays in Getting Civic Actions, 21, may be an easier condition to change. We see by (21,5) that this will have an effect on 5 (Poor Public Services). It is reasonable to expect that if we can improve the ability to get civic action there will be a positive effect on improved public service. Poor Public Services, 5, appears in the block 10 through 28. Therefore, we know that this has an impact on every condition within that block. Looking at the last two rows shows that conditions in this block affect 11 (Alienation of Residents/Difficulty Attracting People), and in turn affect 29 (Difficulty Attracting Investment). Thus, anything such as 21 (Delays in Getting Civic Actions) which affects a variable in this block must also affect our goal of attracting investment.

We must then consider whether improving the ability to obtain civic action contributes positively or negatively to attracting investment. It would seem clear that it does contribute positively. Then we must consider whether the effort spent to improve civic action would be warranted by the size of the effect on attracting investment. If it meets these tests, we add it to our list of possible actions to accomplish our goal.

By this means we would consider a number of possible changes and what their consequences might be. From this we would make a list of actions to be considered further. Note that the matrix itself has not told us what to do, but it has helped by disciplining us to follow the causes and effects that must be considered to develop possible plans of action.

We can also consider the possibilities of changing relations between conditions as well as changing the conditions themselves. Our attention is drawn to High Property Taxes because we can see that it directly affects many other conditions like public services, schools, etc., and these in turn contribute to attracting investment. An interesting relation to look at involving property taxes is the one between 7 (Poor Property Maintenance) and 18 (High Property Taxes). We see that each affects the other—(18,7) and (7,18). Our explanations of (18,7) is that High Property Taxes decrease the money available for maintaining property, a positive effect on Poor Property Maintenance. (Note: This is a positive effect on a condition stated in the negative.) Our explanation of (7,18) is that poor property maintenance yields a lower property value, which implies a lower tax on the property. Given the way these conditions are stated, this is a negative effect. It encourages owners to allow their property to run down, which lowers the value of taxable property, decreasing revenues used to supply public services, schools, etc., which makes it more difficult to attract investment. Thus poor property maintenance is detrimental to our goal.

Seeing this, we are led to ask, What if we change the tax system so that the initial tax on the building is based upon its purchase price, and any improvement thereafter allows the owner to pay a lower tax? This gives an incentive to improve property, which increases revenues, etc. This change would appear to deserve further consideration.

It is often useful to simplify the description of the system by combining similar or closely related conditions to obtain a smaller matrix. We did this before as part of our massaging. We will do it again. In Figure 9.6 we show new numbers assigned to groups of conditions. We combined these to get the matrix in Figure 9.7, giving a new name to the resulting conditions.

Some of the x's in Figure 9.7 have been replaced with numbers. These are level numbers which show marks that may be removed (torn) to reveal the underlying structure within a block. We remove the highest numbered marks from a block first, reorder the rows and columns within that block, remove the next highest numbered marks, etc. (see chapter 3).

The level numbers tell the order in which to remove the marks to reveal the structure within each block. In choosing these numbers, we make four considerations: strength, confidence, immediacy, and binding.

Strength concerns how important the effect is. If the effect is small, we assign a high level number which causes it to be removed from consideration early so we may look at the inner structure involving stronger relations.

We may assign a high number when we lack confidence about the importance or existence of an effect so that it will be removed while looking for the structure involving relations we are more sure of.

We may assign a high number to remove an effect if it is a long-term effect. This allows us to look at the structure within the block associated with more immediate effects.

Shunt diagrams (chapter 4) show where removing a mark or a small number of marks will have a significant effect on reducing the size of the block in the inner system. Thus, we will often tag marks with a high level number to remove them so as to reveal the inner structure. This we call binding.

Although we have four considerations which affect the choice of the order in which the marks are removed, we have only one set of level numbers to represent this choice. We will often make several different orderings using different considerations in choosing level numbers. This is a trial and error process using judgment to find elements to remove to reveal the most meaningful structure. The goal here is to manipulate the matrix so it reveals the clearest understanding.

Figure 9.8 shows the results of reordering the matrix in Figure 9.7. This gives a smaller version of the model in Figure 9.6.

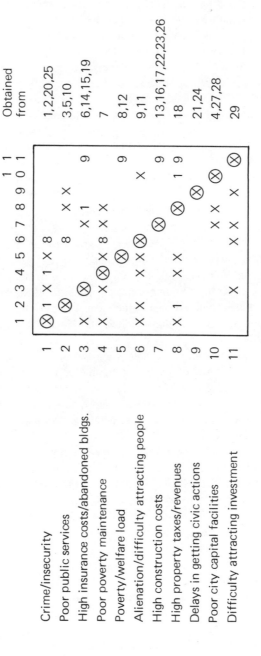

Figure 9.7
Difficulty Attracting Investment: Matrix #7—Matrix #6 Reduced

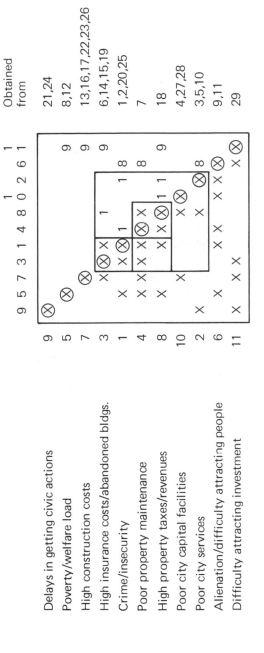

Figure 9.8
Matrix #8—Matrix #7 Reordered

9.7 How to Proceed

We propose that if you were the manager, you could proceed to make a structural model as follows:

1. Define the condition you desire to effect as a consequence of the change you make to the system.
2. List those conditions you believe may affect, either directly or indirectly, either positively or negatively, the condition you wish to effect. (See Table 9.1.)
3. Choose the experts you feel could help you understand these conditions. You may wish to assemble them in one room together, which can be very expensive and noisy, or you may prefer to deal with them individually.
4. Present and discuss with each of the experts the statement of the condition you wish to effect.
5. Get each expert to add to your original list of conditions those the expert feels may influence either directly or indirectly the condition to be effected.
6. Assemble the list of conditions, combining differently stated but equivalent conditions.
7. Distribute this list to the experts for their review, comments, and proposed changes.
8. Get each expert to list for each of his conditions the other conditions which directly affect it. These are the relations between conditions. Try not to list relations that occur only through the intervention of other conditions in the list. For each relation write a brief explanation, giving the reason or mechanism that justifies it. Try to include only the major relations in the initial rounds. Where it is important, code the relations to distinguish for later analysis between significant and less significant ones. (See Table 9.1.)
9. Represent this information as a precedence matrix. (See Figure 9.1.)
10. Reorder the rows and corresponding columns of the matrix to show the structure of the system more clearly. This reordering (a) brings together those conditions that mutually affect each other, and (b) orders the conditions so that the primary causes occur toward the top, and the final effects toward the bottom. (See Figure 9.2)

    This reordering is done using the methods of partitioning and tearing discussed in chapters 2, 3, and 4. The computer program TERABL can be used as an aid in making this reordering.

11. Massage the matrix. Once the matrix is reordered, the cause and effect relations which occur in the initial matrix are displayed so that one can see the structure of the system more clearly. Seeing the structure more clearly, the experts are likely to change their original inputs. In the discussions that occur at this stage the definitions of the conditions may be clarified and perhaps changed. Conditions that are essentially the same may be combined, and irrelevant conditions eliminated. This reduces the size of the matrix and makes it easier to understand. One may wish to reorder the matrix to bring together related conditions to see whether they may be combined. As the matrix becomes massaged the system relations become clearer. As they become clearer it is often noted that further changes are in order. (Figures 9.3 through 9.6 show stages in this massaging.)

12. Look for changes you might be able to make to the system and use the matrix to help you trace through the likely consequences. Changes may be made to conditions or to relations.

13. Evaluate the possible changes, the costs of making the changes, and the costs and benefits of the effects that would occur. The matrix can help you trace the cause and effects of changes you wish to evaluate. The matrix for a cause and effect model will not tell you the nature or size of the effect; these evaluations may require some expert judgment, data collection, and possibly a higher stage of modeling for some critical part of the system, as discussed in the next section.

14. Choose from the alternatives the best course of action.

9.8 Stages of Modeling

Let us consider four stages in the modeling process. In each successive stage more information is put in and more understanding can be gotten out. The stages are as follows:

Structural Model—This is the type of model discussed in this chapter. It is sometimes referred to as a cause and effect model. It can be valuable in itself, or as a basis for the other stages of modeling. Here we list the conditions and/or variables of the system and show which others they cause. This should be done before one develops any model. We have formalized that process so that it can be done more effectively and for larger systems. This form of model is expressed in a precedence matrix. This analysis of structure may yield a desirable result in itself, or be a preliminary step to the later stages of modeling. These later stages we briefly discuss below.

Sign Model—In this stage of modeling we replace the marks in our precedence matrix with $+$, $-$, or $\times$. A plus means a change in one variable or condition produces a change of the same sign in the other. A minus implies that a change in one variable or condition has an effect of the opposite sign in the other. A times sign implies that there are multiple processes that produce mixed effects, of both the same and opposite signs.

The rules for interpreting such a model are quite simple. We trace chains of cause and effect. If the chain contains no times and an even number of minus signs, then a change at the beginning of the chain produces an effect of the same sign at the end of the chain. If there are an odd number of minus signs, the initial change and final effect are of opposite signs. In a circuit, if there is no times, and the number of minuses is even, then a change is reinforcing, while if there are an odd number, the change tends in the direction of self-correction. If the sense of a condition or variable is reversed, e.g., Poor Housing is restated as Good Housing, then all the plus and minus signs in the corresponding row and column of the matrix must be reversed.

Magnitude Model—This model is obtained by replacing each mark in the precedence matrix by the signed derivative of the row variable with respect to the column variable. It shows for a unit change in the column variable the amount of change that occurs in the row variable.

We touched on magnitude models in chapter 8; an input-output model is an example. However, note that if we directly replace the x's in the precedence matrix with derivatives as proposed above, we must place 1's on the diagonal and transpose rows and columns to get the matrix A used in chapter 8. The square of A can be used to compute the magnitude of effects due to chains of length 2, A to the n power for chains of length n. To get the effects due to chains of all lengths, we use the inverse of the matrix $(I - A)$, which equals $I + A + A^2 + A^3 + \ldots$.

Behavioral Model—This form of model need not be a matrix. It may represent each relationship in whatever linear or nonlinear form is required. The Klein-Goldberger model in chapter 8 is an example. In that chapter we demonstrated how it could be useful to show the structure of the system with a precedence matrix even though the model itself might not be a matrix.

TERABL—A Computer Program for Partitioning and Tearing

A.1 General Description of Program

The TERABL computer program was written to (1) assign outputs to a structural matrix as in the procedure represented in Figure 5.9, (2) partition the resulting precedence matrix using level numbers as in procedures 3.2 and 3.4, and (3) generate shunt diagrams for the blocks in the partition as in procedures 4.1 and 4.2.

If one starts with a precedence matrix rather than a structural matrix, the first step will merely assign outputs to the diagonal marks. Thus, the input of a precedence matrix must explicitly include the marks on the diagonal.

TERABL will list the sizes of the blocks of the partition and will print out the matrix before partitioning and the ordered matrix after partitioning. It will print out the shunt diagram for one principal circuit in each block. For each arc in the principal circuit it shows the number of parallel shunts and the lowest index of any parallel shunt. For each vertex in the principal circuit it shows the number of B's and the number of E's in that row.

A.2 Input

The structural, or precedence, matrix is input on File 07. Control of the running of the program is input on File 05.

For each card type the input for the whole card is shown in the form of a Fortran format statement. Then each field is described in sequence.

File 07—The input structural, or precedence, matrix
    First Card—(I3,I3,I3)
        I3: Number of rows (or columns) of the matrix

        I3: Controls printing of input matrix (normally use 2)
          0 or blank: do not print
            1: print on 72-column printer
            2: print on 132-column printer

        I3: Controls printing of output matrix as above (use 2)

    For each row of the matrix
        (I3,A36,I3,5(I3,I3)/(42X,5(I3,I3)))
        I3: Row number. Rows must be numbered and ordered with contiguous numbers from 1 to the total number of rows.

        A36: A description of the row. In the case of a structural matrix for a system of equations this description will be of the variable having the column number corresponding to this row number.

        I3: The number of marks in this row (including the mark on the diagonal for a precedence matrix). For each mark in the row, continued on following cards as required—(I3,I3)
            I3: Level number of the mark (0 to 9). Marks on the diagonal of a precedence matrix must be level number 0.

            I3: Column in which the mark appears.

File 05—To control the running of the program
    This file is made up of sets of records from group A or from group B. These sets can occur in any order provided only that the last record, and *only* the last record, is a STOP.

    Group A—For each pass in which the matrix is partitioned
        Record 1—(A54)
          A54: A title associated with this pass.
        Following records—(26I3)
          I3: Highest level number considered. Higher level numbers are treated as blanks (normally use 9)

I3: Shunt diagrams are generated for all blocks larger than this number (normally use 2)

I3: Maximum index of shunt printed (normally use number of rows in matrix)

I3: Number of marks to be changed. Changes in successive passes are cumulative unless a RESTORE record is used.

For each mark to be changed.
    I3: Row
    I3: Column
    I3: Level number (−1 deletes mark)

Group B—one record beginning in column 1
    "RESTORE": Restores original matrix. Otherwise changes given by group A records are cumulative,
    or
    "STOP": Stops run. Must be last record.

Figure A.1 shows the input for the example in Figure 2.8.

A.3 Output

For each pass corresponding to a set of records from group A we get the following sets of output:

1) The input matrix BEFORE PARTITIONING unless the first record of File 07 specifies that the input matrix is not to be printed. If the matrix will not fit on one page it will be broken up into pages. On each page will be printed PAGE r/c where the pages could be put together so that page r/c is in row r column c. *See Figure A.2a*

b) A heading describing the input from group A which produced this pass, followed by a list for each level of the sizes of the blocks which remain if all marks of higher level number are considered to be blank. See Figure A.2b

c) The matrix AFTER PARTITIONING as ordered by the process of assigning outputs and partitioning as specified in this pass. The outputs occur on the diagonal. (Note: As the partition is redone for each level, only vertices, i.e., rows and columns, within the blocks obtained in the previous partition at the next higher level considered are reordered.) See Figure A.2c

d) For each block in the partition at level zero (i.e., all non-zero marks treated as blank), a shunt diagram is shown for one principal circuit. The size of the block and the length of the principal circuit are shown. The program attempts to pick out one long principal circuit. Column E next to the description gives the original row (i.e., Equation) number corresponding to the vertex, and column V (i.e., Variable) gives the variable assigned to that row in a structural matrix. (For a precedence matrix, E and V are the same.) The index of each shunt is shown above the column where that shunt occurs. The shunts are sorted in increasing order of their index as one goes to the right. Shunts are continued on additional pages as necessary up to the maximum index of shunts indicated on the input. See Figure A.2d & e

e) For each shunt diagram a summary table is presented showing for each vertex in the principal circuit the number of B's (NB) and the number of E's (NE) which line up with that vertex in the shunt diagram. For each arc in the principal circuit on the line between vertices is printed the index of the first parallel shunt (FS), the number of parallel shunts (NS), and also NS-NB and NS-NE. A -1 for FS is printed where there is no parallel shunt. Tearing the arc represented by a line drawn here looks good if FS is equal to -1 (meaning there is no shunt), or if FS is large (the blocks remaining after this tear are likely to be small), or if NS is small (the number of shunts to be torn or avoided because they are not parallel to tears in the principal circuit is small). NS-NB is the subtraction from NS of the number of these shunts which begin on the previous vertex in the principal circuit. This gives the number of shunts which would remain if the arc in the principal circuit corresponding to a line here and all the parallel arcs which begin with a B on the previous vertex are torn. If this number is small and significantly lower than NS, then one might consider tearing the shunt arcs exiting the previous vertex. These tears would all occur in the same column in the precedence matrix. NS-NE is the subtraction from NS of the number of these shunts which end on the following vertex in the principal circuit. This gives the number of shunts which would remain if the arc in the principal circuit corresponding to this line and all the parallel arcs which end with an E on the next vertex are torn. If this number is small and significantly lower than NS, then one might consider tearing the shunt arcs entering the following vertex in the principal circuit. These tears would all occur in the same row in the precedence matrix. See Figure A.2f

Figure A.2 shows the output corresponding to the input shown in Figure A.1. See section 8.7 for a discussion of the tearing of this example.

```
$       DATA    07
$       INCODE  IBMF
20   2  2
  1      C         CONSUM EXP          5      1      3      6     10     17
  2      I         CAPTL FORM          1      2
  3      S P       CORP SAV            2      3      4
  4      P C       CORP.PROFIT  .   - - - - · · 2   4     17
  5      D         CAP CONSUM          3      5     16     19
  6      W 1       EMP COMP            3      6      5     16
  7      N W       NO. WAGE ERN        4      7      5     16·    19
  8        W       WAGE INDX           2      8      7
  9      F I       IMPORTS             5      9      6     10     18     19
 10      A 1       FARM INCOME         6     10      3      6     11     17
 10                                          18
 11        P A     AG PRICES INDX      2     11     18
 12      L 1       PER ASSETS          6     12      3      6     10     14
 12                                          17
 13      L 2       ENT ASSETS          4     13      6     15     18
 14      I L       BOND YIELD          1     14
 15      I S       COMM YIELD          1     15
 16      Y         NAT INCOME          5     16      1      2      5      9
 17      P         OTH INCOME          4     17      6     10     16
 18        P       PRICE INDX          4     18      6      7      8
 19      K         PRIV CAP            3     19      2      5
 20      B         CORP SURPLS         2     20      3
$       DATA    05
$       INCODE  IBMF
KLEIN-GOLDBERGER MODEL
  9  2 20   0
LEVEL 9
  9  2 20   4   5 16   9   6 16   9   7 16   9 17 16   9
LEVEL 8
  9  2 20   3   5 19   8 10 17   8 10   3   8
STOP
```

Figure A.1
Input to TERABL Program—Klein-Goldberger Model

```
KLEIN-GOLDBERGER MODEL                              BEFORE PARTITIONING
   PAGE  1/ 1
                                    0 0 0 0 0 0 0 0 0 0 0 0 0 0 0 0 0 0 ·0 ·0 0
                                    0 0 0 0 0 0 0 0 0 1 1 1 1 1 1 1 1 1 1 2
                                    1 2 3 4 5 6 7 8 9 0 1 2 3 4 5 6 7 8 9 0
        C       CONSUM EXP       1  X . X . . X . . . X . . . . . . X . . .
        I       CAPTL FORM       2  . X . . . . . . . . . . . . . . . . . .
        S P     CORP SAV         3  . . X X . . . . . . . . . . . . . . . .
        P C     CORP PROFIT      4  . . . X . . . . . . . . . . . . . X . .
        D       CAP CONSUM       5  . . . . X . . . . . . . . . . . X . . X .
        W 1     EMP COMP         6  . . . . X X . . . . . . . . . . X . . .
        N W     NO. WAGE ERN     7  . . . . X . X . . . . . . . . . X . . X .
          W     WAGE INDX        8  . . . . . . X X . . . . . . . . . . . .
        F I     IMPORTS          9  . . . . . X . . X X . . . . . . . X X .
        A 1     FARM INCOME     10  . . X . . X . . . X X . . . . . X X . .
          P A   AG PRICES INDX  11  . . . . . . . . . . X . . . . . . X . .
        L 1     PER ASSETS      12  . . X . . X . . . X . X . X . . X . . .
        L 2     ENT ASSETS      13  . . . . . X . . . . . X . X . X . X . .
        I L     BOND YIELD      14  . . . . . . . . . . . . . X . . . . . .
        I S     COMM YIELD      15  . . . . . . . . . . . . . . X . . . . .
        Y       NAT INCOME      16  X X . . X . . X . . . . . . . X . . . .
        P       OTH INCOME      17  . . . . . X . . X . . X . . . . X X . .
          P     PRICE INDX      18  . . . . . X X X . . . . . . . . . X . .
        K       PRIV CAP        19  . X . . X . . X . . . . . . . . . . X .
        B       CORP SURPLS     20  . . X . . . . . . . . . . . . . . . . X
```

Figure A.2a
Output from TERABL Program—Klein-Goldberger Model

```
KLEIN-GOLDBERGER MODEL
ORDER   20
HIGHEST LEVEL MARK TO BE CONSIDERED    0
LARGEST BLOCK FOR WHICH SHUNT DIAGRAM IS NOT PRINTED    2
LARGEST INDEX SHUNT TO BE CONSIDERED  20


LEVEL    0
          1  14   1   1   1   1   1
```

Figure A.2b

```
KLEIN-GOLDBERGER MODEL                                              AFTER PARTITIONING
    PAGE   1/ 1

                                    0 0 0 0 0 0 0 0 0 0 0 0 0 0 0 0 0 0 0 0
                                    0 0 0 0 0 0 0 0 0 1 1 1 1 1 1 1 1 1 1 2
                                    2 1 3 4 5 6 7 8 9 0 1 6 7 8 9 4 2 5 3 0
    I          CAPTL FORM       2  X . . . . . . . . . . . . . . . . . . .
    C          CONSUM EXP       1  . X X . . . X . . . . X . . X . . . . .
    S P        CORP SAV         3  . . X X . X . . . . X . . X . . . . . .
    P C        CORP PROFIT      4  . . . X . . . . . . . . . X . . X . . .
    D          CAP CONSUM       5  . . . . X . . . . X . . X . . X . . . .
    W 1        EMP COMP         6  . . . . X X . . . . . X . . . . . . . .
    N W        NO. WAGE ERN     7  . . . . X . X . . X . . X . . X . . . .
    W          WAGE INDX        8  . . . . . . X X . . . . . . . . . . . .
    F I        IMPORTS          9  . . . . . X . . X X . . . X X . . . . .
    A 1        FARM INCOME     10  . . X . . X . . X X . X X . X X . . . .
    P A        AG PRICES INDX  11  . . . . . . . . . . X . . X . . X . . .
    Y          NAT INCOME      16  X X . . X . . . X . . X . . X . . . . .
    P          OTH INCOME      17  . . . . X . . . X X . X X . . . . . . .
    P          PRICE INDX      18  . . . . . X X X . . . . . X . . . . . .
    K          PRIV CAP        19  X . . . X . . . . . . . . . X . . . . .
    I L        BOND YIELD      14  . . . . . . . . . . . . . . X . . . . .
    L 1        PER ASSETS      12  . . X . . X . . . X . . X . . X X . . .
    I S        COMM YIELD      15  . . . . . . . . . . . . . . . . X . . .
    L 2        ENT ASSETS      13  . . . . . X . . . . . . X . . . X X . .
    B          CORP SURPLS     20  . . X . . . . . . . . . . . . . . . . X
```

Figure A.2c

```
    BLOCK SIZE   14
SHUNT DIAGRAM
PRINCIPAL CIRCUIT LENGTH   12
VARIABLE NAME                           E    V    1    1    1    2
        Y          NAT INCOME          16   16
        D          CAP CONSUM           5    5              B
        K          PRIV CAP            19   19              I
        N W        NO. WAGE ERN         7    7         B    E
            W      WAGE INDX            8    8         1
            P      PRICE INDX          18   18    B    E
            P A    AG PRICES INDX      11   11    I
        A 1        FARM INCOME         10   10    E
        P          OTH INCOME          17   17                   B
        P C        CORP PROFIT          4    4                   I
        S P        CORP SAV             3    3                   I
        C          CONSUM EXP           1    1                   E
```

Figure A.2d

```
BLOCK SIZE   14
SHUNT DIAGRAM
PRINCIPAL CIRCUIT LENGTH   12
VARIABLE NAME
```

		E	V	X 10	X 10	10
Y	NAT INCOME	16	16	I	I	E
D	CAP CONSUM	5	5	I	E	B
K	PRIV CAP	19	19	I	B	I
N W	NO. WAGE ERN	7	7	I	I	I
W	WAGE INDX	8	8	I	I	I
P	PRICE INDX	18	18	I	I	I
P A	AG PRICES INDX	11	11	I	I	I
A 1	FARM INCOME	10	10	E	I	I
P	OTH INCOME	17	17	B	I	I
P C	CORP PROFIT	4	4	I	I	I
S P	CORP SAV	3	3	I	I	I
C	CONSUM EXP	1	1	I	I	I

Figure A.2e

Shunt Diagram in Figure A2.d continued

2	2 (7,16)	3	3 (6,16)	3	4	5	5 (6,16)	5	6 (6,16)	7 (17,16)	8	8 (X)	8	8	9 (6,16)
	B		B	E			B	E	B	B		1	E	E	B
B	I		I		B	B	I		I	I	B	I	B		I
I	I		I		I	I	I		I	I	I	1	I	B	I
I	E		I		I	I	I		I	I	I	I	I	I	I
6			I		I	I	I		I	I	I	I	I	I	I
E			E		I	I	I	B	I	I	I	I	I	I	I
					6	I	6	I	I	I	I	I	I	I	I
		B		B	E	6	E	I	6	I	I	E	I	I	I
		I		I		E		I	E	E	I		1	I	I
		I		I				I			I		1	I	I
		I		I				I			6		B	6	I	6
		E		9				9			E		I	9	9	E

```
  BLOCK SIZE   14
SHUNT DIAGRAM SUMMARY
  23 SHUNTS
```

VARIABLE NAME			E	V	NB	NE	FS	NS	NS-NB	NS-NE
Y		NAT INCOME	16	16	6	5	2	9	③	8—
D		CAP CONSUM	5	5	7	1	1	15	8	15
K		PRIV CAP	19	19	2	0	1	17	15	15
N W		NO. WAGE ERN	7	7	1	2	1	16	15	16
	W	WAGE INDX	8	8	0	0	1	16	16	13
	P	PRICE INDX	18	18	2	3	1	15	13	15
	P A	AG PRICES INDX	11	11	0	0	1	15	15	10
A 1		FARM INCOME	10	10	2	5	3	12	10	9
P		OTH INCOME	17	17	2	3	2	11	9	11
P C		CORP PROFIT	4	4	0	0	2	11	11	11
S P		CORP SAV	3	3	1	0	2	12	11	8
C		CONSUM EXP	1	1	0	4	3	8	8	3

Figure A.2f

```
LEVEL 9                                                    BEFORE PARTITIONING
    PAGE   1/  1
                                         0 0 0 0 0 0 0 0 0 0 0 0 0 0 0 0 0 0 0 0·
                                         0 0 0 0 0 0 0 0 0 1 1 1 1 1 1 1 1 1 1 2
                                         1 2 3 4 5 6 7 8 9 0 1 2 3 4 5 6 7 8 9 0
      C        CONSUM EXP        1  X . X . . X . . . X . . . . . . . X . . .
      I        CAPTL FORM        2  . X . . . X . . . X . . . . . . . . . . .
      S P      CORP SAV          3  . . X X . . . . . . . . . . . . . . . . .
      P C      CORP PROFIT       4  . . . X . . . . . . . . . . . . . X . . .
      D        CAP CONSUM        5  . . . . X . . . . . . . . . . . 9 . . X .
      W 1      EMP COMP          6  . . . X X . . . . . . . . . . . 9 . . . .
      N W      NO. WAGE ERN      7  . . . X . X . . . . . . . . . . 9 . . X .
      W        WAGE INDX         8  . . . . . . X X . . . . . . . . . . . . .
      F I      IMPORTS           9  . . . . X . . X X . . . . . . . . . X X .
      A 1      FARM INCOME      10  . . X . . X . . . X X . . . . . . X X . .
      P A      AG PRICES INDX   11  . . . . . . . . . . X . . . . . . Y . . .
      L 1      PER ASSETS       12  . . X . . X . . . X . X . X . . X . . . .
      L 2      ENT ASSETS       13  . . . . . X . . . . . . X . X . X . . X .
      I L      BOND YIELD       14  . . . . . . . . . . . . . X . . . . . . .
      I S      COMM YIELD       15  . . . . . . . . . . . . . . X . . . . . .
      Y        NAT INCOME       16  X X . . X . . X . . X . . . . . . X . . .
      P        OTH INCOME       17  . . . . . X . . . X . . . X . . . 9 X . . .
      P        PRICE INDX       18  . . . . . X X X . . . . . . . . . . X . .
      K        PRIV CAP         19  . X . . X . . . . . . . . . . . . . X .
      B        CORP SURPLS      20  . . X . . . . . . . . . . . . . . . . . X
```

Figure A.2g

```
LEVEL 9
ORDER   20
HIGHEST LEVEL MARK TO BE CONSIDERED    9
LARGEST BLOCK FOR WHICH SHUNT DIAGRAM IS NOT PRINTED    2
LARGEST INDEX SHUNT TO BE CONSIDERED   20
CHANGES   ROW COL SEN
           5  16   9
           6  16   9
           7  16   9
          17  16   9


LEVEL    9
        1 14  1   1   1   1   1


LEVEL    0
        1  2  1   1   1   1   1  4  1   1   1   1   1   1   1   1   1
```

Figure A.2h

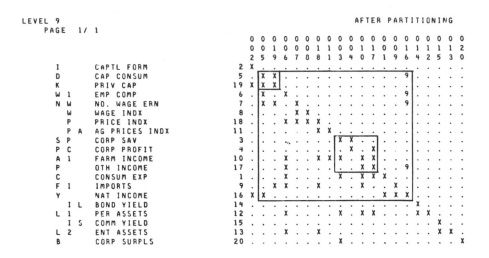

Figure A.2i

```
   BLOCK SIZE   4
SHUNT DIAGRAM
PRINCIPAL CIRCUIT LENGTH    4
VARIABLE NAME                          E    V    2
      A  1     FARM INCOME            10   10    E
      P        OTH INCOME             17   17    B
      P  C     CORP PROFIT             4    4    I
      S  P     CORP SAV                3    3    I
```

Figure A.2j

```
   BLOCK SIZE   4
SHUNT DIAGRAM SUMMARY
   1 SHUNTS
```

VARIABLE NAME			E	V	NB	NE	FS	NS	NS-NB	NS-NE
A 1	FARM INCOME		10	10	0	1	-1	0	0	0
P	OTH INCOME		17	17	1	0	2	1	0	1
P C	CORP PROFIT		4	4	0	0	2	1	1	1
S P	CORP SAV		3	3	0	0	2	1	1	0 —

Figure A.2k

```
LEVEL 8                                                    BEFORE PARTITIONING
    PAGE   1/ 1
                                     0 0 0 0 0 0 0 0 0 0 0 0 0 0 0 0 0 0 0 0
                                     0 0 0 0 0 0 0 0 0 1 1 1 1 1 1 1 1 1 1 2
                                     1 2 3 4 5 6 7 8 9 0 1 2 3 4 5 6 7 8 9 0
        C       CONSUM EXP      1  X . X . . X . . . X . . . . . . X . . .
        I       CAPTL FORM      2  . X . . . . . . . . . . . . . . . . . .
        S P     CORP SAV        3  . . X X . . . . . . . . . . . . . . . .
        P C     CORP PROFIT     4  . . X . . . . . . . . . . . . . X . . .
        D       CAP CONSUM      5  . . . . X . . . . . . . . . . . 9 . . 8 .
        W 1     EMP COMP        6  . . . . X X . . . . . . . . . . 9 . . . .
        N W     NO. WAGE ERN    7  . . . X . X . . . . . . . . . . 9 . . X .
          W     WAGE INDX       8  . . . . . . X X . . . . . . . . . . . .
        F I     IMPORTS         9  . . . . . X . . X X . . . . . . . . X X .
        A 1     FARM INCOME    10  . . 8 . . X . . . X X . . . . . 8 X . .
        P A     AG PRICES INDX 11  . . . . . . . . . . X . . . . . . X . .
        L 1     PER ASSETS     12  . . X . . X . . . X . X . X . . X . . .
        L 2     ENT ASSETS     13  . . . . . X . . . . . . X . X . . X . .
        I L     BOND YIELD     14  . . . . . . . . . . . . . X . . . . . .
        I S     COMM YIELD     15  . . . . . . . . . . . . . . X . . . . .
        Y       NAT INCOME     16  X X . . X . . . X . . . . . . X . . . .
        P       OTH INCOME     17  . . . . X . . . X . . . . . . 9 X . . .
          P     PRICE INDX     18  . . . . . X X X . . . . . . . . . X . .
        K       PRIV CAP       19  X . . X . . . . . . . . . . . . . . X .
        B       CORP SURPLS    20  . . X . . . . . . . . . . . . . . . . X
```

Figure A.2l

```
LEVEL 8
ORDER   20
HIGHEST LEVEL MARK TO BE CONSIDERED   9
LARGEST BLOCK FOR WHICH SHUNT DIAGRAM IS NOT PRINTED   2
LARGEST INDEX SHUNT TO BE CONSIDERED  20
CHANGES   ROW COL SEN
              5  16   9
              6  16   9
              7  16   9
             17  16   9
              5  19   8
             10  17   8
             10   3   8
```

```
LEVEL   9
        1 14  1  1  1  1  1
```

```
LEVEL   8
        1  2  1  1  1  )  1  4  1  1  1  1  1  1  1  1  1
```

```
LEVEL   0
        1  1  1  1  1  1  1  1  1  1  1  1  1  1  1  1  1  1  1  1
```

Figure A.2m

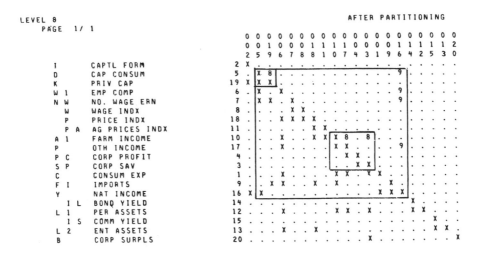

Figure A.2n

Prior Literature

Several researchers have used graphs to represent systems of equations. Harary (1960, 1962) derived his graph from the permutation group describing the determinant for a linear system of equations. Mason (1953), independently in the electrical engineering literature, derived his graph from the determinant solution of linear equations and the analogy to what he calls signal flow in electrical networks. Tribus (1958) used diagrams to represent the effect of variables on equations. Parter (1961) used graphs to show the effect of fill-in during elimination.

Harary (1962) developed a technique for partitioning matrices which required the same permutations of the rows and the columns. This limitation has since been overcome by the device of assigning an output to each equation. Two papers introducing this device were published independently in the same month—Dulmage and Mendelsohn (1962) who built upon Harary's framework, and Steward (1962) who built upon the concepts of information flow.

Runyon proposed the problem of tearing a graph subject to a priori criteria such as a minimum number of arcs or a minimum sum of weights of arcs to obtain a directed graph with no circuits [Seshu and Reed:61]. Younger's thesis (1963) and an unpublished paper by Woolmansee (1968) deal with this problem. Sargent and Westerberg (1964) approached this problem using dynamic programming techniques. We do not pretend to solve this problem, nor has anyone else provided a complete, efficient solution to it.

The problem we have solved is somewhat different. Our concern is with tearing, subject to technical considerations evaluated by the user. For large problems this becomes an interactive process between the user and the computer. Formulated this way, useful solutions are obtained more easily.

Kron (1963) developed a technique he called tearing for constructing the solution of large, primarily linear, systems from the solutions of subsystems [Bueckner:56, Noble:69, Steward:65b]. The amount of work required to construct the solution from the pieces depends upon where it is torn. The general applicability of Kron's technique has been severely limited by the lack of practitioners with Kron's skill in choosing where to tear. This was discussed in chapter 5.

On the other hand, the techniques developed here have been taught in a few hours of lecture to students in the General Electric Advanced Engineering Course. These students have then applied these techniques to engineering problems which they encountered in their work assignments. In addition, our techniques are applicable to systems which are nonlinear, and even to human decision processes where in a formal sense nothing other than precedence relations exist describing the order in which variables may be determined. Partitioning and tearing have been applied to systems of several hundred variables.

This book is a natural consequence of earlier papers that the author published on partitioning and tearing. More precise definitions and careful proofs have been added to the results of these papers, and many new results have been developed.

Warfield and associates [Warfield: 76, Baldwin: 75, Malone: 75], initially at Battelle Memorial Institute in Columbus, have developed and applied techniques related to those presented here to a variety of societal systems. Chapter 9 is a melding of their techniques and those developed independently by the author.

Bibliography

Alexander, C.W.
1964 *Notes on the Synthesis of Form*. Cambridge, Mass.: Harvard University Press.

Alexander uses non-directed graphs to indicate variables which are directly related. His technique does not distinguish between sets of variables which may be determined one at a time sequentially and those which must be determined iteratively. He uses a clustering technique based upon a criterion justified by information theory. His techniques are well known in the fields of architecture and urban planning. Alexander's methods motivated ideas in structured programming where they gained greater success than they had in their original domain. (See section 2.9 of this book.)

Ando, A.; Fisher, F.M.; and Simon, H.A.
1963 *Essays on the Structure of Social Science Models*. Cambridge, Mass.: MIT Press.

A collection of articles by these authors on the subject of nearly decomposable systems.

Ando, A., and Modigliani, F.
1969 "Econometric Analysis of Stabilization Policies." *Papers and Proceedings of the American Economic Association* 59:296–314.

They analyze the FRB-MIT model by using actual time paths for all

271

variables outside those of a submodel being studied. Step by step, they expand the submodel by making more of these exogenous variables endogenous. This relates to the use of tearing discussed in chapter 8.

Ashby, W.R.
1960 *Design for a Brain.* New York: John Wiley & Sons.
Ashby makes a probability argument to show that evolutionary adaptation is more likely to be successful by a series of small changes than by more radical change.

Baldwin, Maynard M. (ed)
1975 *Portraits of Complexity: Applications of Systems Methodologies to Societal Problems,* The Battelle Monograph Series.

Berge, C.
1958 *The Theory of Graphs.* London: Dunrod.
Our graph theory definitions largely follow Berge. Chapter 10, "Matching of a Simple Graph," considers the assignment problems which we see as the problem of assigning an output set.

Blalock, H.M., Jr.
1967 *Causal Inferences in Non-Experimental Research.* Chapel Hill, N.C.: University of North Carolina Press.

Bohm, Corrado, and Jacopini, Guiseppe
1966 "Flow Diagrams, Turing Machines and Languages with Only Two Formation Rules." *Communications of the ACM* 9 (May): 366–371.

Brillouin, L.
1956 *Science and Information Theory.* New York: Academic Press, Inc.
Introduces the concept that the utility of a model can be measured by how much it reduces the a priori uncertainty in the prediction.

Bueckner, H.F.
1956 *Remarks on Diacoptics.* Report DF-56TG104, Technical Information Series, General Electric Co., Schenectady, N.Y.
Bueckner represents Kron's technique in matrix form. The discussion of Kron's method in chapter 5 follows this paper.

Churchman, C. West
1968 *The Systems Approach.* New York: Delta Publishing Co.

Deo, N.
1969 *An Extensive English Language Bibliography on Graph Theory and Its Applications.* Report 32-1413, Jet Propulsion Laboratory, California Institute of Technology, Pasadena, Calif. (October) (Supplement April 1971)
In this bibliography of over 3,000 entries, there is not much evidence

of recognizable contributions to the feedback tear set problem other than Younger:63. This appears to be a reflection of the state of the art rather than any gap in Deo's interest since Deo himself has been concerned with this problem (Deo:70).

Deo, N.
1970 "Cycle Elimination in Weighted Digraphs by Orientation Reversal of Edges." *Proceedings of The Third Hawaii International Conference on System Sciences,* University of Hawaii, Honolulu (January):419–422.
Shows the equivalence between a feedback reversal set and a feedback tear set. Deo shows a heuristic based upon only local properties, the in and out degrees of the vertices. Also shows a way to enumerate all feedback cut sets.

Deosoer, C.A.
1960 "The Optimum Formula for the Gain of a Flow Graph in a Simple Derivation of Coate's Formula." *Proceedings of the IRE* 48 (May):883–889.
Good discussion of Mason's theory of signal flow graphs. See Mason:53 and 56.

Dorfman, Robert; Samuelson, Paul A.; and Solow, Robert M.
1958 *Linear Programming and Econometric Models.* New York: McGraw-Hill.
Discussion of Input/Output models.

Duesenberry, J.S.; Fromm, G.; Klein, J.R.; and Kuh, E. (eds)
1965 *The Brookings Quarterly Econometric Model of the United States.* Chicago: Rand McNally & Company.
Partitioning was used as a tool in solving this system.

Duesenberry, J.S., and Klein, J.R. (cds)
1965 "Introduction: The Research Strategy and Its Application." In *The Brookings Quarterly Econometric Model of the United States,* edited by Duesenberry, J.S., Fromm, G., Klein, L.R., and Kuh, E. Chicago: Rand McNally & Company.

Dulmage, A.L., and Mendelsohn, N.S.
1962 "On the Inversion of Sparse Matrices." *Mathematics of Computation* 16 (October):494–496.
Generalizes the result of Harary:62 to different permutations of rows and columns, i.e., QAP, by use of an assignment algorithm. This result is equivalent to the result of Steward:62 except their result was justified only in the case of linear systems. Both papers were published in the same month. They introduced the use of bipartite graphs.

Eisner, M.
1969 *Guide for the TROLL System*. Report, MIT (June).
TROLL is an interactive system for studying econometric models which includes a partitioning operation.

Fisher, F.M.
1965 "Dynamic Structure and Estimation in Economy-Wide Econometric Models." In *The Brookings Quarterly Econometric Model of the United States*, edited by Duesenberry, J.S., Fromm, G., Klein, L.R., and Kuh, E. Chicago: Rand McNally & Company.

Frazer, R.A.; Duncan, W.J.; and Collar, A.R.
1953 *Elementary Matrices*. Cambridge, Mass.: Cambridge University Press.

Garner, W.R.
1962 *Uncertainty and Structure as Psychological Concepts*. New York: John Wiley & Sons.
Applies Brillouin's concept of the utility of a model to contingency tables.

Hall, M.
1956 "An Algorithm for Distinct Representatives." *American Mathematical Monthly* 63:716–717.

Hall, P.
1935 "On Representatives of Subsets." *Journal of the London Mathematical Society* 10:26–30.

Harary, F.
1959–60 "A Graph Theoretic Method for the Complete Reduction of a Matrix with a View Toward Finding its Eigenvalues." *Journal of Mathematical Physics* 38:104–111.
Applies a symmetry transformation, i.e., $P^{-1}AP$, to bring a matrix to block triangular form. This transformation preserves the eigenvalues. The set of eigenvalues of A is the union of eigenvalues of the matrices which appear as blocks on the diagonal of $P^{-1}AP$. Harary derives an adjacency matrix from the permutation group which arises from the determinant. He gets transitive closure by taking the nth power of the adjacency matrix.

Harary, F.
1962 "A Graph Theoretic Approach to Matrix Inversion by Partitioning." *Numerische Mathematik* 4:128–135.
Applies the theory of Harary:59–60 to matrix inversion, i.e., $(P^{-1}AP)^{-1}$. It is shown in Dulmage and Mendelsohn:62 and Stew-

ard:62 that by solving an assignment problem, this can be generalized to $(QAP)^{-1}$, i.e., different permutations of the rows and
columns.

Himmelblau, D.H., and Bischoff, K.B.
1967 *Process Analysis and Simulation*. New York: John Wiley & Sons.
An extensive discussion is presented on the use of partitioning in
chemical process design.

Holt, C.C.; Shirey, R.; Steward, D.V.; Midler, J.L.; and Stroud, A.H.
1964 *Program SIMULATE, A User's and Programmer's Manual*. Report,
University of Wisconsin, Social Systems Research Institute (May).
(mimeographed)
This program is used to study the simulation of econometric models,
perform sensitivity analyses, etc. It includes a partitioning
algorithm.

Holt, C.C.
1965 "Validation and Application of Macroeconomic Models Using
Computer Simulation." In *The Brookings Quarterly Econometric
Model of the United States*, edited by Duesenberry, J.S., Fromm,
G., Klein, L.R., and Kuh, E. Chicago: Rand McNally.
Makes a case for the importance of analyzing the structure and
causal chains of an econometric model as a guide in experimentation
on the model and calculation of the model.

Hoggatt, A.C., and Balderston, F.E. (eds)
1963 "Simulation Models: Analytic Variety and the Problem of Model
Reduction." In *Symposium on Simulation Models: Methodology and
Applications to the Behavioral Sciences*. Cincinnati: South-Western
Publishing Co.
Makes a case for the reduction of models in order to diagnose and
improve them.

Householder, A.S.
1953 *Principles of Numerical Analysis*. New York: McGraw-Hill.
Refers to papers by Sherman and Morrison (1949, 1950), Woodbury
(1950), and Bartlett (1951), which give formulas for the inverse of a
modified matrix. These formulas are related to Kron's method. See
Householder:60.

Householder, A.S.
1960 "Matrix Inversion." In *The International Dictionary of Applied
Mathematics*. New York: Van Nostrand Co.
Makes the same observation regarding Kron's method as was made
by Bueckner:56.

Hymans, S.H., and Shapiro, H.T.
1970 *The DHL-III Quarterly Econometric Model of the U.S. Economy.*
Report, Research Seminar in Quantitative Econometrics, University
of Michigan, Ann Arbor, Michigan.

Kelly, J.E., Jr.
1961 "Critical Path Planning and Scheduling, Mathematical Basis."
Operations Research 9 (May–June):296–320.

Klein, L.R., and Goldberger, A.S.
1955 *An Econometric Model of the United States, 1929–1952.* Amsterdam:
North-Holland Publishing Co.

Krolak, P.; Felts, W.; and Marble, G.
1971 "A Man-Machine Approach Toward Solving the Traveling Salesman
Problem." *Communications of the ACM* 14 (May).
The authors make a case for the use of man-machine interaction in
applying heuristics to obtain satisfactory answers to a difficult
combinatorial problem.

Kron, G.
1953 "A Set of Principles to Interconnect the Solutions of Physical
Systems." *Journal of Applied Physics* 24 (August):965–980.

Kron, G.
1963 *Diakoptics.* London: MacDonald.

Leontief, W.
1965 "The Structure of the U.S. Economy." *Scientific American* 212
(April):25–35.

Lorens, C.S.
1964 *Flowgraphs for the Modeling and Analysis of Linear Systems.* New
York: McGraw-Hill.
A good short book on signal flow theory. Indicates that the loop
theorem of signal flow theory was first put forth by Claude Shannon
in a classified report in the early 1940s.

McCarthy, M.D.
1972 *The Wharton Quarterly Econometric Forecasting Model—Mark III.*
Philadelphia: Wharton School of Finance and Commerce.

Malone, David W.
1975 "An Introduction to the Applications of Interpretive Structural
Modeling." Article in Baldwin:75.

Marschak, J.
1950 "Statistical Inference in Economics: An Introduction." *Statistical
Inference in Dynamic Economic Models,* Cowles Commission

Monograph 10, edited by T.C. Koopmans. New York: John Wiley & Sons.

Mason, S.J.
1953 "Feedback Theory—Some Properties of Signal Flow Graphs." *Proceedings of the IRE* 41 (September):1,144–56.
Presents a technique for solving linear systems of equations based upon a topology derived from the equations for electrical circuits and using the techniques of determinants. Some of the concepts are similar to the techniques developed in Steward:62. Although an output assignment is assumed, this assumption is never explicitly stated.

Mason, S.J.
1956 "Feedback Theory—Further Properties of Signal Flow Graphs." *Proceedings of the Institute of Radio Engineers* 44 (July):920–926
See Mason:53.

Miller, George A.
1956 "The Magical Number Seven, Plus or Minus Two: Some Limitations on Our Capacity for Processing Information." *Psychological Review* 63:81–97.

Munroe, I.
1969 *Efficient Determination of the Strongly Connected Components and Transitive Closure of a Directed Graph.* Report, Department of Computer Science, University of Toronto.

Noble, B.
1969 *Applied Linear Algebra.* Englewood Cliffs, N.J.: Prentice-Hall.

Orcutt, G.H.
1952 "Actions, Consequences, and Causal Relations." *The Review of Economics and Statistics* 34:305–314.
Uses the concept that if B always happens when A happens, then it may be said to "cause" B. If it also occurs that A always happens when B does, then it may also be said that B causes A. This concept is then extended to correlations.

Parter, S.
1961 "The Use of Linear Graphs in Gauss Elimination." *SIAM Review* 3:119–130.
Uses graph theory to consider fill-in during elimination.

Purdom, P.W.
1968 *A Transitive Closure Algorithm.* Report #33, Computer Science Department, University of Wisconsin.

Rudd, D.F., and Steward, D.V.
1964 "On Information Flow and Process Design Calculations." Presented to the American Institute of Chemical Engineers, Las Vegas (September).
Introduces the use of tearing into the analysis of process systems with feedback.

Saaty, T.L.
1959 *Mathematical Methods of Operations Research.* New York: McGraw-Hill.

Sargent, R.W.H., and Westerberg, A.W.
1964 "Speed-Up in Chemical Engineering Design." *Transactions of the Institution of Chemical Engineers,* 42:T190–T197.
Proposes finding strongly connected subgraphs (partitioning) by tracing circuits. Also applies dynamic programming to finding the feedback tear set.

Sage, Andrew P.
1977a *Methodology for Large-Scale Systems.* New York: McGraw-Hill.

Sage, Andrew P., editor
1977b *Systems Engineering: Methodology and Applications.* New York: IEEE Press.

Seshu, S., and Reed, M.B.
1961 *Linear Graphs and Electrical Networks.* Reading Mass.: Addison-Wesley.

Sherman, J., and Morrison, W.J.
1950 "Adjustments of an Inverse Matrix Corresponding to a Change in One Element of a Given Matrix." *Annals of Mathematical Statistics* 21:124.

Simon, H.A.
1957 *Models of Man: Social and Rational.* New York: John Wiley & Sons.
Chapter I: Causal Ordering and Identifiability," pp. 10–36.
Chapter II: "Spurious Correlation: A Causal Interpretation," pp.37–49.
Chapter III: "On the Definition of the Causal Relation," pp.50–61.

Simon, H.A.
1969 *The Sciences of the Artificial.* Cambridge, Mass.: MIT Press.

Steward, D.V.
1959 "On an Algebraic Foundation for Constructing Optimum Algorithms." *Preprints of the 14th National Meeting of the ACM* (September).

Steward, D.V.
1962 "On an Approach to Techniques for the Analysis of the Structure of Large Systems of Equations." *SIAM Review* 4 (October):321–342.

Steward, D.V.
1965a *Partitioning and Tearing Systems of Equations.* Report #581, Mathematics Research Center, United States Army, The University of Wisconsin (June).

Steward, D.V.
1965b "Partitioning and Tearing Systems of Equations." *Journal of SIAM, Numerical Analysis,* Series E, vol. 2.

Steward, D.V.
1965c "Partitioning and Tearing Large Systems of Equations." In *Proceedings of the Symposium on Systems Theory.* Brooklyn: Polytechnic Press of the Polytechnic Institute of Brooklyn, pp. 89–95.

Steward, D.V.
1966a *An Approach to the Simulation of Some Classes of Micro Models.* Systems Formulation, Methodology, and Policy Workshop, Paper 6608, University of Wisconsin (June).

Steward, D.V.
1966b *The Analysis of the Sensitivity of Models.* Systems Formulation, Methodology, and Policy Workshop, Paper 6602, Social Systems Research Institute, University of Wisconsin (January).

Steward, D.V.
1966c *The Simulation and Analysis of Micro Models.* Systems Formulation, Methodology, and Policy Workshop, Paper 6610, Social Systems Research Institute, University of Wisconsin (July).

Steward, D.V.
1967a *Critical Path Scheduling: Introduction, Critique, and Extensions.* Report APED-5525, Technical Publications Unit, Atomic Power Equipment Department, General Electric Co., San José, Calif.

Steward, D.V.
1967b *An Improved Method for Tearing Large Systems.* Report APED-5526, Technical Publications Unit, Atomic Power Equipment Department, General Electric Co., San José, Calif.

Steward, D.V.
1968 *The Design Structure System.* Report APED-5538, Technical Publications Unit, Atomic Power Equipment Department, General Electric Co., San José, Calif.

Steward, D.V.
1969 "Tearing Analysis of the Structure of Disorderly Sparse Matrices." In *Proceedings of the Symposium on Sparse Matrices and Their Applications,* edited by Willoughby (Report RA 1 3-12-69), Thomas J. Watson Research Center, I.B.M., Yorktown Heights, N.Y., pp. 65–74.

Strotz, R.H., and Wold, H.A.O.
1960 "Recursive vs. Nonrecursive Systems: An Attempt at Synthesis." *Econometrica* 28 (April):417–427.

Suits, D.B.
1962 "Forecasting and Analysis with an Econometric Model." *American Economic Review* (March):104–132.

Tribus, M.
1958 *On an Aid to System Optimization.* Report #58GL236, General Engineering Laboratory, General Electric Co.

Warfield, John N.
1976 *Societal Systems: Planning, Policy and Complexity.* New York: John Wiley & Sons.
Warfield and his associates at Battelle Memorial Institute, and several contributors to the IEEE Transactions on Systems, Man and Cybernetics, have built a literature on modeling techniques related to those developed in chapter 9. They use the term Interpretive Structural Modeling. See also Sage:77a, 77b and Baldwin:75.

Weil, R.L., Jr., and Steward, D.V.
1967 "The Question of Determinancy in Square Systems of Equations." *Zeitschrift Fur Nationalökonomie* 27:261–266.

Wold, H.A.O.
1959 "Ends and Means in Econometric Model Building." In *Probability and Statistics, The Harold Cramer Volume,* edited by U. Grenander. New York: John Wiley & Sons.

Wold, H.A.O.
1960 "A Generalization of Causal Chain Models." *Econometrica* 28 (April):443–463.

Wold, H.A.O., and Jureen, L.
1953 *Demand Analysis, A Study in Econometrics.* New York: John Wiley & Sons.

Woodbury, Max
1950 *Inverting Modified Matrices.* Memorandum Report 42, Statistical Research Group, Princeton, N.J.

Woodmansee, G.H.
1968 *Results Related to the Determination of Minimum Feedback Arc Sets for a Directed Graph*. Term paper, Computer Sciences/Math 837—Graph Theory, Fall 1967–68, University of Wisconsin.

Younger, D.H.
1963 *Feedback in a Directed Graph*. Ph.D. Thesis, Columbia University.

Index

Legend
  p = preface
  f = following
  underscore = definition